Lecture Notes in Mathematics

Edited by A. Dold and B. Eckmann

1203

Stochastic Processes and Their Applications

Proceedings of the International Conference
held in Nagoya, July 2–6, 1985

Edited by K. Itô and T. Hida

Springer-Verlag
Berlin Heidelberg New York London Paris Tokyo

Editors

Kiyosi Itô
Research Institute for Mathematical Sciences
Kyoto University
Kyoto, Japan 606

Takeyuki Hida
Department of Mathematics, Faculty of Science
Nagoya University
Nagoya, Japan 464

Mathematics Subject Classification (1980): 60, 62, 35J, 35K; 70, 81, 82, 92

ISBN 3-540-16773-0 Springer-Verlag Berlin Heidelberg New York
ISBN 0-387-16773-0 Springer-Verlag New York Berlin Heidelberg

Printed in Germany

Printing and binding: Beltz Offsetdruck, Hemsbach/Bergstr.
2146/3140-543210

PREFACE

The Fifteenth Conference on Stochastic Processes and Their Applications was held in Nagoya, Japan, for the period July 2-6, 1985. This volume contains the invited papers presented at this conference.

The conference was attended by 360 scientists, from all over the world, and 110 among them were participants from overseas, whose attendance is greatly appreciated.

In addition to the invited paper sessions, there were contributed paper sessions on the following topics.

Gaussian processes and fields
Branching, population and biological models
Stochastic methods in physics
Stochastic differential equations
Probability distributions and limit theorems
Stable processes and measures
Random walks and i.i.d. random variables
Filtering, control and optimization
Statistical inference
Diffusion processes
Markov processes
Storage and reliability
Ergodic theorems
Martingales and processes with independent increments
Point processes and applications
Stochastic processes in random media

The organizers regret that the papers in these sessions could not be included in this volume.

We should like to express sincere thanks to Professors Ken-iti Sato and Tadahisa Funaki who have helped us in editing this volume.

September 15, 1985

Kiyosi Itô

Takeyuki Hida

TABLE OF CONTENTS

The following lectures were delivered at the conference, but are not included in this volume.

G. Kallianpur

Some recent results in nonlinear filtering theory with finitely additive white noise

H. Kesten

First-passage percolation and maximal flows

S. Orey

On getting to a goal fast

Yu.A. Rozanov

On stochastic partial differential equations

A.S. Sznitman

A "propagation of chaos" result for Burger´s equation

The following special lecture was also given.

K. Itô

An outline of the development of stochastic processes.

ASYMPTOTIC BEHAVIOUR OF STOCHASTIC FLOWS OF DIFFEOMORPHISMS

Peter H. Baxendale
Department of Mathematics
University of Aberdeen
Aberdeen AB9 2TY
Scotland

1. Introduction

Let M be a compact smooth manifold of dimension d. Consider the following stochastic differential equation on M:

$$d\xi_t(x) = V_0(\xi_t(x))dt + \sum_{i\geq 1} V_i(\xi_t(x)) \circ dW_t^i. \tag{1.1}$$

Here V_i, $i \geq 0$, are smooth vector fields on M and $\{W_t^i : t \geq 0\}$, $i \geq 1$, are independent real-valued Brownian motions on some probability space $(\Omega, \mathfrak{F}, P)$. We write $\circ\, dW_t^i$ to denote the Stratonovich stochastic differential. By convention we shall always take $\xi_0(x) = x$.

Infinite dimensional noise is allowed in equation (1.1). In this case we impose the condition that $\sum_{i\geq 1} X_i V_i$ converges in $C^\infty(TM)$ almost surely whenever the X_i are independent N(0,1) random variables. Here $C^\infty(TM)$ denotes the space of smooth vector fields on M with the topology of uniform C^r convergence for all $r \geq 0$. Then $W_t = \sum_{i\geq 1} W_t^i V_i$ exists for all $t \geq 0$ and is a Wiener process with values in $C^\infty(TM)$ corresponding to an abstract Wiener space $(i, H, C^\infty(TM))$ for some real separable Hilbert space H and continuous inclusion $i : H \hookrightarrow C^\infty(TM)$. Equation (1.1) may now be written more concisely as

$$d\xi_t(x) = V_0(\xi_t(x)) + \circ\, dW_t(\xi_t(x)). \tag{1.2}$$

Conversely any real separable Hilbert space H continuously included in $C^\infty(TM)$ yields an abstract Wiener space and a Wiener process $\{W_t : t \geq 0\}$. We may realise $W_t = \sum_{i\geq 1} W_t^i V_i$ by choosing $\{V_i : i \geq 1\}$ to be a complete orthonormal basis of H. See [1], section 2, for details.

For fixed $x \in M$ the process $\{\xi_t(x) : t \geq 0\}$ is a diffusion process on M whose infinitesimal generator is given on C^2 functions by

$$L = V_0 + \frac{1}{2} \sum_{i\geq 1} V_i^2. \tag{1.3}$$

If we consider the dependence of $\xi_t(x)$ on the initial position $x \in M$ we obtain a random family of maps $\xi_t : M \to M$ for $t \geq 0$. It is well established that by choosing suitable versions of $\{\xi_t(x) : t \geq 0\}$ for each $x \in M$ we obtain a random process $\{\xi_t : t \geq 0\}$ with values in Diff(M), the group of smooth diffeomorphisms of M with the topology of uniform C^r convergence for all $r \geq 0$. The process $\{\xi_t : t \geq 0\}$ is called a stochastic flow (of diffeomorphisms). For results on the existence of the stochastic flow of diffeomorphisms for equation (1.1) see [6], [12], [15] or [19].

The process $\{\xi_t : t \geq 0\}$ has the following properties

(i) independent increments on the left (i.e. if $0 \leq t_0 < t_1 < \ldots < t_n$ then $\xi_{t_i}\xi_{t_{i-1}}^{-1}$, $1 \leq i \leq n$, are independent)

(ii) time homogeneous (i.e. if $t > s$ the distribution of $\xi_t\xi_s^{-1}$ depends only on $t-s$)

(iii) continuous sample paths with probability 1 (w.p.1)

(iv) $\xi_0 = I$ w.p.1.

It is shown in [2] that any process $\{\xi_t : t \geq 0\}$ in Diff(M) satisfying (i) - (iv) arises as the solution of an equation of the form (1.1). See also [23]. The law of the process $\{\xi_t : t \geq 0\}$ is determined uniquely by the drift $V_0 \in C^\infty(TM)$ and the Hilbert space H. Since H is continuously included in $C^\infty(TM)$ it is a reproducing kernel Hilbert space of sections of TM and so it is determined by a positive semi-definite reproducing kernel b. We have

$$\begin{aligned} b(x,y) &= E(W_1(x) \otimes W_1(y)) \\ &= \sum_{i \geq 1} V_i(x) \otimes V_i(y) \qquad (1.4) \\ &\in T_xM \otimes T_yM \end{aligned}$$

for $x,y \in M$. V_0 and b may be interpreted as the mean and variance of the random vector field which is driving (1.1). Notice that the individual V_i, $i \geq 1$, determine the law of the flow only so far as they contribute to the sum (1.4). It is important to distinguish between the behaviour of the one-point motion $\{\xi_t(x) : t \geq 0\}$, which is characterised by L, and the behaviour of the flow $\{\xi_t : t \geq 0\}$. The extra information given by V_0 and b which is not contained in L is the correlation between the one-point motions $\{\xi_t(x) : t \geq 0\}$ for different x but the same noise $\{W_t^i : t \geq 0\}$, $i \geq 1$.

In this paper we shall be concerned with the following questions about the stochastic flow $\{\xi_t : t \geq 0\}$.

(1) <u>Geometrical nature of ξ_t</u>. Does there exist a nice subset D of Diff(M) such that $\xi_t \in D$ for all $t \geq 0$ w.p.1? Since the stochastic flow has continuous sample paths we can always take D equal to the

identity component $\mathrm{Diff}_I(M)$ of Diff(M). Of more interest is the case where G is a subgroup of $GL(d,\mathbb{R})$ and D is the group of automorphisms of some G-structure on M. See Kobayashi [18] for details on G-structures. For example D might be the group of isometries of some Riemannian structure on M or the group of diffeomorphisms which preserve some smooth volume element on M.

(2) Stability of solutions. For distinct $x,y \in M$ what happens to $d(\xi_t(x),\xi_t(y))$ as $t \to \infty$? (Here d denotes the metric corresponding to some Riemannian structure on M.) This is a question about the 2-point motion $\{(\xi_t(x),\xi_t(y)) : t \geq 0\}$ on M^2. We obtain a similar question, but one which involves infinitely many points, by considering $\mathrm{diam}(\xi_t(U))$ for some neighbourhood U of x in M.

(3) Induced measures. Let $\mathcal{P}(M)$ denote the space of Borel probability measures on M. For $\rho \in \mathcal{P}(M)$ let $\rho_t = \rho\xi_t^{-1}$. Then $\{\rho_t : t \geq 0\}$ is a Markov process in $\mathcal{P}(M)$. In [22] Le Jan obtains results on the nature of the stationary distribution for this Markov process. See also [20]. We shall say $\rho \in \mathcal{P}(M)$ is invariant under the stochastic flow if $\rho_t = \rho$ for all $t \geq 0$ w.p.1. In this case we may take D in question 1 to be the group of ρ-preserving diffeomorphisms of M. This is a much stronger property of ρ than the fact that ρ is a stationary measure for the one-point process; in fact ρ is stationary for the one-point process if and only if $E(\rho_t) = \rho$ for all $t \geq 0$. Since the one-point process is a Feller process and M is compact there exists at least one stationary probability measure. Moreover if L is elliptic then ρ is unique and the one-point process is ergodic with respect to ρ. Henceforth we shall use ρ to denote a stationary probability measure on M. In general ρ will not be invariant under the flow and we wish to describe the behaviour of ρ_t as $t \to \infty$.

We shall describe some general results in Sections 3 and 4. Before that we consider two specific examples of stochastic flows on the circle; in each case we may write down an explicit solution. In the final section we consider a family of examples of stochastic flows on the torus.

For results for similar questions applied to isotropic stochastic flows in Euclidean spaces see Baxendale and Harris [5] and Le Jan [21].

2. Two stochastic flows on the circle

Before proceeding with general results we consider the following examples of stochastic flows. In each case $M = S^1 = \mathbb{R}/2\pi\mathbb{Z}$ with the Riemannian structure inherited from the standard inner product on $\mathbb{R}$. We write θ for elements of both $\mathbb{R}$ and S^1 and observe that equations (2.1) and (2.3) below have period 2π.

<u>Example 1</u>

$$d\xi_t(\theta) = dW_t. \tag{2.1}$$

The solution is given by

$$\xi_t(\theta) = \theta + W_t \pmod{2\pi}, \tag{2.2}$$

that is, ξ_t is rotation of S^1 through the angle W_t.

<u>Example 2</u>

$$d\xi_t(\theta) = \sin(\xi_t(\theta)) \circ dW_t^1 + \cos(\xi_t(\theta)) \circ dW_t^2. \tag{2.3}$$

This is the equation for the gradient stochastic flow obtained by taking the standard embedding of S^1 as the unit circle in $\mathbb{R}^2$. See Carverhill, Chappell and Elworthy [10] (which also contains a computer simulation of the 10-point motion due to P. Townsend and D. Williams). The solution is given by

$$\tan(\tfrac{1}{2}\xi_t(\theta)) = \frac{\tan(\frac{1}{2}U_t^2) + a_t(\tan(\frac{1}{2}\theta) - b_t)}{1 - a_t(\tan(\frac{1}{2}\theta) - b_t)\tan(\frac{1}{2}U_t^2)} \tag{2.4}$$

where

$$a_t = \exp(-U_t^1 + \tfrac{1}{2}t) \tag{2.5}$$

$$b_t = \int_0^t a_s^{-1} dU_s^2 \tag{2.6}$$

and $\{U_t^i : t \geqslant 0\}$, $i = 1,2$, are Brownian motions given by

$$\begin{bmatrix} dU_t^1 \\ dU_t^2 \end{bmatrix} = R_t \begin{bmatrix} dW_t^1 \\ dW_t^2 \end{bmatrix}$$

where R_t denotes rotation in $\mathbb{R}^2$ through the angle $\xi_t(\pi)$.

In both examples the one-point motion has generator given by $L = \frac{1}{2}\frac{d^2}{d\theta^2}$, so that the one-point motion is Brownian motion on S^1. The stationary probability measure ρ is normalised Lebesgue measure. In example 1 the stochastic flow consists of rotations, whereas in example 2 the stochastic flow lies in a 3-dimensional Lie subgroup (isomorphic to $SL(2,\mathbb{R})$) of $\mathrm{Diff}(S^1)$. In example 1 distances between points are preserved and the measure ρ is invariant under the flow.

In example 2 the fact that $a_t \to \infty$ as $t \to \infty$ w.p.1 implies that distances and the measure ρ become more and more distorted by ξ_t as $t \to \infty$. More precisely b_t converges to some random value b_∞, say, as $t \to \infty$ w.p.1, and if θ_1 and θ_2 are distinct points of S^1 with $\tan(\frac{1}{2}\theta_i) \neq b_\infty$, $i = 1,2$, then

$$\lim_{t\to\infty} \frac{1}{t} \log d(\xi_t(\theta_1), \xi_t(\theta_2)) = \lim_{t\to\infty} \frac{1}{t} \log\left(\frac{1}{a_t}\right)$$

$$= -\frac{1}{2}.$$

Also the measure ρ_t becomes more and more concentrated but does not converge as $t \to \infty$. Instead it looks more and more like a unit mass attached to a Brownian particle. In fact for any $\theta \in S^1$ such that $\tan(\frac{1}{2}\theta) \neq b_\infty$ then $\rho_t - \delta(\xi_t(\theta)) \to 0$ weakly as $t \to \infty$ (where $\delta(\theta)$ denotes the unit mass concentrated at θ). For a generalisation of example 2 to gradient stochastic flows on spheres see [3].

3. The support theorem for stochastic flows

Let C denote the space of continuous functions $f : [0,\infty) \to \mathrm{Diff}(M)$ such that $f(0) = I$, with the compact-open topology. By properties (iii) and (iv) of stochastic flows, $\xi = \{\xi_t : t \geq 0\}$ is a C-valued random variable. In this section we study ν, the distribution of ξ. The following theorem is due to Ikeda and Watanabe [16].

Theorem 3.1. Let $\mathcal{U}$ denote the space of piecewise constant functions $u : [0,\infty) \to H$. Let $\xi^u \in C$ denote the solution of

$$\frac{d}{dt}(\xi_t^u(x)) = (V_0 + u(t))(\xi_t^u(x))$$

$$\xi_0^u(x) = x.$$

Then the support of ν is the closure in C of $\{\xi^u : u \in \mathcal{U}\}$.

Theorem 3.1 can sometimes provide an answer to question 1. Let LA(H) denote the closed Lie subalgebra of $C^\infty(TM)$ generated by H (or, equivalently, generated by $\{V_i : i \geq 1\}$). Notice that LA(H) depends only on H as a set and not on its inner-product. Theorem 3.1 remains valid if we replace H by LA(H) in the definition of $\mathcal{U}$. It follows that if $LA(H) = C^\infty(TM)$ then the support of ν is C itself, and so the only closed subset D of Diff(M) satisfying $\xi_t \in D$ for all $t \geq 0$ w.p.1 is $D = \mathrm{Diff}_I(M)$. On the other hand if the (deterministic) flows along the individual vector fields V_i, $i \geq 0$, all lie in some closed subgroup D of Diff(M) then $\xi_t \in D$ for all $t \geq 0$ w.p.1. The stochastic flows studied in the previous section provide examples of this phenomenon. For an example where D is the group of conformal diffeomorphisms of a sphere see [3]. In general, even if the noise in (1.1) is only finite-dimensional the resulting stochastic flow is infinite-dimensional.

When considering the behaviour of ξ_t as $t \to \infty$ the support theorem is often of little use. This is because the topology on C is that of uniform convergence on compact subsets of $[0,\infty)$ and so, for example, the set

$$\{f \in C : d(f(t)(x), f(t)(y)) \to 0 \text{ as } t \to \infty\},$$

for fixed x and y in M, is not closed in C. The failure of the support of ν to provide information about ν itself is shown up in the following result.

Theorem 3.2. Let ν and ν' denote the distributions in C of the stochastic flows corresponding to pairs (V_0, H) and (V_0', H').

(i) If $LA(H) = LA(H')$ and $V_0' - V_0 \in LA(H)$ then ν and ν' have the same support.

(ii) Either ν and ν' are singular or $\nu = \nu'$. The latter case occurs if and only if $V_0 = V_0'$ and $H = H'$ (as Hilbert spaces).

Proof. (i) This follows directly from Theorem 3.1.

(ii) For $T > 0$ let $C_T = \{f|_{[0,T]} : f \in C\}$ and let ν_T denote the distribution of $\{\xi_t : 0 \leq t \leq T\}$ in C_T. Consider the isomorphism of C with $(C_T)^\infty$ given by

$$f_n(t) = f(t+nT)(f(nT))^{-1}, \quad n \geq 0,\ t \in [0,T],\ f \in C.$$

Under this isomorphism ν corresponds to the infinite product of copies of ν_T. The first part of (ii) now follows from a theorem of Kakutani [17] on infinite product measures. The second part is contained in [2]. □

4. Lyapunov exponents for stochastic flows

Consider the derivative flow $\{D\xi_t : t \geq 0\}$ on the tangent bundle TM of M. For each $x \in M$ and $t \geq 0$, $D\xi_t(x)$ is a random linear mapping from T_xM to $T_{\xi_t(x)}M$. In this section we shall consider the limiting rate of growth of $D\xi_t(x)$ as $t \to \infty$ for $x \in M$, and its effect on the nature of the stochastic flow as $t \to \infty$. If we impose a Riemannian structure we may consider $(D\xi_t(x)^* D\xi_t(x))^{1/2} : T_xM \to T_xM$, the positive part of $D\xi_t(x)$. Here $D\xi_t(x)^* : T_{\xi_t(x)} \to T_xM$ denotes the adjoint of $D\xi_t(x)$. The following theorem goes back to results of Furstenberg

[13] on products of independent identically distributed random matrices and Oseledec [24] on products of a stationary ergodic sequence of matrices. It was adapted to apply to deterministic flows by Ruelle [25] and its present formulation for stochastic flows is due to Carverhill [7].

Theorem 4.1. Assume ρ is a stationary ergodic probability measure for the one-point process of a stochastic flow $(\xi_t : t \geq 0)$ defined on a probability space $(\Omega, \mathcal{F}, P)$. For $P \times \rho$-almost all $(\omega, x) \in \Omega \times M$,

$$(D\xi_t(x)^* D\xi_t(x))^{1/2t} \to \Lambda_{(\omega,x)} \text{ as } t \to \infty \tag{4.1}$$

where $\Lambda_{(\omega,x)}$ is a random linear map on T_xM with non-random eigenvalues

$$e^{\lambda_1} \geq e^{\lambda_2} \geq \ldots \geq e^{\lambda_d} > 0. \tag{4.2}$$

The values $\lambda_1 \geq \lambda_2 \geq \ldots \geq \lambda_d$ are called the Lyapunov exponents for the stochastic flow. Since M is compact, any two Riemannian structures on M are uniformly equivalent and so the Lyapunov exponents are independent of the choice of Riemannian structure. They are non-random because they depend only on the remote future of the stochastic flow, whereas $\Lambda_{(\omega,x)}$ is in general random because the eigenspaces corresponding to distinct eigenvalues depend on the entire evolution of the stochastic flow. Roughly speaking, the theorem implies that the positive part of $D\xi_t(x)$ has eigenvalues growing like $e^{\lambda_1 t}, e^{\lambda_2 t}, \ldots, e^{\lambda_d t}$ as $t \to \infty$. More precisely we have:

Corollary 4.2.

(i) For ρ-almost all $x \in M$ and Lebesgue-almost all $v \in T_xM \setminus \{0\}$

$$P\{\tfrac{1}{t} \log \|D\xi_t(x)(v)\| \to \lambda_1 \text{ as } t \to \infty\} = 1. \tag{4.3}$$

(ii) For ρ-almost all $x \in M$

$$P\{\tfrac{1}{t} \log \det(D\xi_t(x)) \to \lambda_1 + \lambda_2 + \ldots + \lambda_d \text{ as } t \to \infty\} = 1. \tag{4.4}$$

We comment that if the generator L for the one-point process is elliptic then we may remove the condition on x in Corollary 4.2. If the generator for the induced one-point process $\{D\xi_t(x)(v) : t \geq 0\}$ in TM is elliptic on $\bigcup_{y \in M} (T_yM \setminus \{0\})$ then we may remove the condition on $v \in T_xM \setminus \{0\}$ also.

Corollary 4.2(i) provides an answer to the infinitesimal version of question 2. Carverhill [7], following Ruelle [25], has a local stable manifold theorem which enables us to obtain answers to question 2 as originally posed. We state a special case of the theorem.

Theorem 4.3. Suppose $\lambda_1 < 0$. If $0 > \mu > \lambda_1$ then for $P \times \rho$-almost all (ω,x) in $\Omega\times M$ there exist (measurable) $r(\omega,x) > 0$ and $\gamma(\omega,x) > 0$ such that $d(x,y_i) < r(\omega,x)$ for $i = 1,2$ implies

$$d(\xi_t(y_1),\xi_t(y_2)) < \gamma(\omega,x)d(y_1,y_2)e^{\mu t} \tag{4.5}$$

for all $t \geq 0$.

An immediate consequence of Theorem 4.3 is that if $\lambda_1 < 0$ then for ρ-almost all $x \in M$ and all $\varepsilon > 0$ there exists $\delta > 0$ such that

$$P\{\text{diam } \xi_t(B(x,\delta)) \to 0 \text{ as } t \to \infty\} > 1 - \varepsilon \tag{4.6}$$

where $B(x,\delta)$ denotes the open ball with centre x and radius δ.

Let us review example 2. Let $\overline{\theta}(\omega)$ be the random point of S^1 given by $\tan(\frac{1}{2}\overline{\theta}(\omega)) = b_\infty$. Since the distribution of $\{\xi_t : t \geq 0\}$ is rotation invariant it follows that $\overline{\theta}(\omega)$ is uniformly distributed on S^1. The set of full $P\times\rho$ measure on which (4.1) is valid is the set $\{(\omega,\theta) : \theta \neq \overline{\theta}(\omega)\}$, and $\lambda_1 = -\frac{1}{2}$. In Theorem 4.3 we need $r(\omega,\theta) < d(\theta,\overline{\theta}(\omega))$. In (4.6) notice that diam $\xi_t(B(\theta,\delta)) \to 0$ as $t \to \infty$ whenever $\overline{\theta}(\omega) \in S^1 - B(\theta,\delta)$, and this happens with probability $1 - \frac{\delta}{\pi}$. So corresponding to $\varepsilon > 0$ we need $\delta < \pi\varepsilon$.

The following result provides a global version of Corollary 4.2(ii), so long as ρ is closely related to the Riemannian measure m, say, on M. Let $\lambda_\Sigma = \lambda_1 + \lambda_2 + \ldots + \lambda_d$.

<u>Theorem 4.4</u>. Suppose ρ has a positive C^2 density with respect to m. Then

$$\lim_{t\to\infty} \frac{1}{t} I(\rho_t|\rho) = -\lambda_\Sigma \qquad \text{w.p.1}$$

where $I(\rho_t|\rho)$ denotes the relative entropy of ρ_t with respect to ρ. In particular $\lambda_\Sigma \leq 0$, and $\lambda_\Sigma = 0$ if and only if ρ is invariant under the stochastic flow $\{\xi_t : t \geq 0\}$.

<u>Proof</u>. See Baxendale [4], Theorem 5.2.

The highest Lyapunov exponent λ_1 may be evaluated using a formula due to Carverhill [8], following the method of Has'minskii [14] for linear stochastic differential equations. We shall obtain a special case of the formula in the next section. The formula may be generalised so as to obtain the sum of the k highest Lyapunov exponents for $1 \leq k \leq d$. See Baxendale [4] for details. The sum λ_Σ has been studied by Chappell [11].

For a recent survey article on Lyapunov exponents for stochastic flows see Carverhill [9].

5. A family of stochastic flows on the torus

Let $M = (\mathbb{R}/2\pi\mathbb{Z})^2 = T^2$, the two-dimensional torus, with coordinates $x = (x^1,x^2) \in M$ and Riemannian structure inherited from the standard inner product on $\mathbb{R}^2$. Consider the one-parameter family of stochastic flows on M, parametrised by $\alpha \in [0,\pi/2]$, given by

$$d\xi_t(x) = \sum_{i=1}^{4} V_i(\xi_t(x)) \circ dW_t^i \tag{5.1}$$

where the vector fields are as follows

$$V_1(x) = \sin x^1 (\cos\alpha \frac{\partial}{\partial x^1} + \sin\alpha \frac{\partial}{\partial x^2})$$

$$V_2(x) = \cos x^1 (\cos\alpha \frac{\partial}{\partial x^1} + \sin\alpha \frac{\partial}{\partial x^2})$$

$$V_3(x) = \sin x^2 (-\sin\alpha \frac{\partial}{\partial x^1} + \cos\alpha \frac{\partial}{\partial x^2})$$

$$V_4(x) = \cos x^2 (-\sin\alpha \frac{\partial}{\partial x^1} + \cos\alpha \frac{\partial}{\partial x^2}).$$

For each α the infinitesimal generator for the one-point motion is one half of the Laplace-Beltrami operator on M, so that the one-point motion is Brownian motion on M. The stationary probability measure ρ for the one-point motion is normalized Lebesgue measure. It is easily checked by calculating the covariance $b(x,y)$ that different values of $\alpha \in [0,\pi/2]$ give rise to stochastic flows with different distributions.

We may simulate the k-point motion $\{(\xi_t(x_1),\dots,\xi_t(x_k)) : t \geq 0\}$ in M^k for any $k \geq 1$, $(x_1,\dots,x_k) \in M^k$ and α using the scheme

$$\xi_{nh+h}(x_r) = \xi_{nh}(x_r) + \sqrt{h} \sum_{i=1}^{4} X_i^{(n)} V_i(\xi_{nh}(x_r))$$

for $1 \leq r \leq k$ and $n \geq 0$, where $\{X_i^{(n)} : 1 \leq i \leq 4,\ n \geq 0\}$ are independent $N(0,1)$ random variables. Figures 1 to 4 show the result of simulating with $h = 0{\cdot}01$ the 225-point motion started from the regular 15×15 lattice J in M given by $J = \{(\frac{(2i-1)\pi}{15}, \frac{(2j-1)\pi}{15}) : 1 \leq i,j \leq 15\}$. We show $[0,2\pi] \times [0,2\pi]$ squares whose edges are to be identified to give M. The pictures correspond to different values of α but they are all at the same time $t = 0{\cdot}5$ and are all generated by the same sequence of $X_i^{(n)}$. In order to appreciate better the distortion of M caused by ξ_t we have added in figures 2, 3 and 4 the straight line segments joining the current positions of nearest neighbour pairs in the original lattice J.

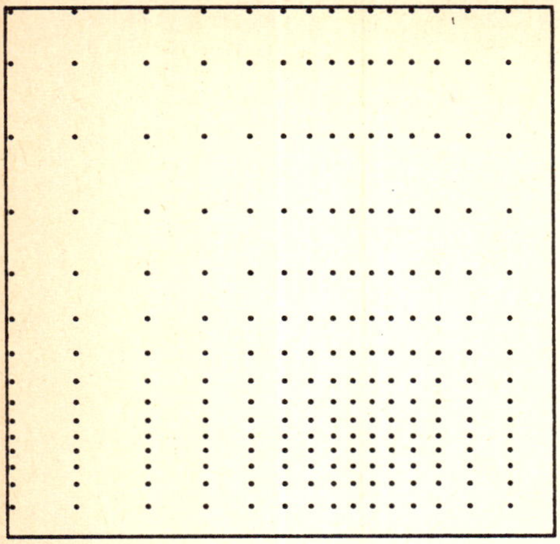

Figure 1 : $\alpha = 0$, $t = 0 \cdot 5$

Figure 2 : $\alpha = \pi/6$, $t = 0 \cdot 5$

Figure 3 : $\alpha = \pi/3$, $t = 0 \cdot 5$

Figure 4 : $\alpha = \pi/2$, $t = 0 \cdot 5$

Consider the two extreme values of α. If $\alpha = 0$ then we obtain

$$\xi_t(x) = (\xi_t^1(x^1), \xi_t^2(x^2))$$

where $\{\xi_t^1 : t \geqslant 0\}$ and $\{\xi_t^2 : t \geqslant 0\}$ are independent copies of the stochastic flow on S^1 studied in example 2. This decomposition shows up in Figure 1 in the way in which horizontal lines are mapped to horizontal lines and vertical lines are mapped to vertical lines. The stochastic flow takes values in a 6-dimensional Lie subgroup (isomorphic to $SL(2,\mathbb{R}) \times SL(2,\mathbb{R})$) of Diff(M). This is the only value of α for which ξ_t takes values almost-surely in a finite-dimensional subgroup of Diff(M) because in all other cases the Lie algebra generated by $\{V_1,V_2,V_3,V_4\}$ is infinite-dimensional. From the calculations in example 2 we see that for $x,y \in M$ with $x \neq y$

$$P\{\lim_{t\to\infty} \frac{1}{t} \log d(\xi_t(x),\xi_t(y)) = -\frac{1}{2}\} = 1$$

and that for $x \in M$

$$P\{\text{weak-}\lim_{t\to\infty} (\rho_t - \delta(\xi_t(x))) = 0\} = 1.$$

If $\alpha = \pi/2$ then we obtain the stochastic flow studied by Ikeda and Watanabe in [16]. In this case div $V_i = 0$ for $1 \leqslant i \leqslant 4$ so that ρ is invariant under the stochastic flow. See [16] for an exact characterisation of the support of the stochastic flow. For $x \neq y$ the two-point process $\{(\xi_t(x),\xi_t(y)) : t \geqslant 0\}$ on $M^2 \setminus D$ (where D denotes the diagonal $\{(u,v) \in M^2 : u = v\}$) has finite stationary probability measure $\rho \times \rho$. Hence the process $\{(\xi_t(x),\xi_t(y)) : t \geqslant 0\}$ on $M^2 \setminus D$ is recurrent on $M^2 \setminus D$ and so

$$P\{d(\xi_t(x)),\xi_t(y)) \to 0 \text{ as } t \to \infty\} = 0.$$

From the above discussion we have $\lambda_1 = \lambda_2 = -\frac{1}{2}$ when $\alpha = 0$, and $\lambda_1 + \lambda_2 = 0$ (so that $\lambda_1 = -\lambda_2 \geqslant 0$) when $\alpha = \pi/2$. We proceed to calculate λ_1 and λ_2 for general α. We shall write $\lambda_i(\alpha)$ to denote the dependence of λ_i on α. In view of Theorems 4.3 and 4.4 we wish to discover when $\lambda_1(\alpha) < 0$ and when $\lambda_1(\alpha) + \lambda_2(\alpha) = 0$.

We may treat (5.1) as an equation in $\mathbb{R}^2$. Converting to Itô form we obtain

$$d\xi_t(x) = \sum_{i=1}^{4} V_i(\xi_t(x))dW_t^i$$

since in this case the Itô correction term is identically zero. Differentiating with respect to x and writing $D\xi_t(x) = A_t$, we obtain

$$dA_t = \sum_{i=1}^{4} DV_i(\xi_t(x))A_t dW_t^i. \tag{5.2}$$

Calculating $DV_i(y)$ for $1 \leqslant i \leqslant 4$ and substituting in (5.2) we obtain

$$dA_t = \begin{bmatrix} \cos\alpha & 0 \\ \sin\alpha & 0 \end{bmatrix} A_t dU_t^1 + \begin{bmatrix} 0 & -\sin\alpha \\ 0 & \cos\alpha \end{bmatrix} A_t dU_t^2 \tag{5.3}$$

where

$$\left.\begin{aligned} dU_t^1 &= (\cos\xi_t^1(x))dW_t^1 - (\sin\xi_t^1(x))dW_t^2 \\ dU_t^2 &= (\cos\xi_t^2(x))dW_t^3 - (\sin\xi_t^2(x))dW_t^4 \end{aligned}\right\} \tag{5.4}$$

and $\xi_t(x) = (\xi_t^1(x),\xi_t^2(x))$. From (5.4) we see by Lévy's theorem that if $U_0^1 = U_0^2 = 0$ then $\{U_t^1 : t \geq 0\}$ and $\{U_t^2 : t \geq 0\}$ are Brownian motion processes. It follows from (5.3) that for this family of examples the law of $\{D\xi_t(x) : t \geq 0\}$ for fixed $x \in M$ is independent of x. Fix $v \in T_xM \cong \mathbb{R}^2$ with $v \neq 0$ and let $v_t = D\xi_t(x)(v) = A_t v$. Then

$$dv_t = \begin{bmatrix} \cos\alpha & 0 \\ \sin\alpha & 0 \end{bmatrix} v_t dU_t^1 + \begin{bmatrix} 0 & -\sin\alpha \\ 0 & \cos\alpha \end{bmatrix} v_t dU_t^2. \tag{5.5}$$

Write $v_t = r_t(\cos\theta_t, \sin\theta_t)$. Then by Itô's lemma we obtain from (5.5)

$$\begin{aligned} d\theta_t = & -\cos\theta_t \sin(\theta_t - \alpha)dU_t^1 + \sin\theta_t \cos(\theta_t - \alpha)dU_t^2 \\ & + \tfrac{1}{2}(\sin(2\theta_t - \alpha)\cos(2\theta_t - \alpha) - \sin\alpha\cos\alpha)\, dt \end{aligned}$$

and

$$d(\log r_t) = \cos\theta_t \cos(\theta_t - \alpha)dU_t^1 + \sin\theta_t \sin(\theta_t - \alpha)dU_t^2 + q_\alpha(\theta_t)dt$$

where

$$q_\alpha(\theta) = \tfrac{1}{2}\{-1 + \sin^2\alpha + \sin^2(2\theta - \alpha)\}. \tag{5.6}$$

Therefore $\{\theta_t : t \geq 0\}$ is a diffusion on S^1 with infinitesimal generator given by

$$L_1 = \tfrac{1}{4}(\sin^2\alpha + \sin^2(2\theta - \alpha))\frac{d^2}{d\theta^2} + \tfrac{1}{2}(\sin(2\theta - \alpha)\cos(2\theta - \alpha) - \sin\alpha\cos\alpha)\frac{d}{d\theta}$$

and so for $0 < \alpha \leq \frac{\pi}{2}$ it has a unique stationary probability measure μ_α, say, with smooth positive density g_α, say, with respect to Lebesgue measure. Also

$$\log r_t = \log r_0 + M_t + \int_0^t q_\alpha(\theta_s)ds$$

where M_t is a martingale with $\frac{d}{dt}\langle M\rangle_t \leq 1$. Therefore

$$\begin{aligned} \lim_{t\to\infty} \frac{1}{t}\log r_t &= \lim_{t\to\infty} \frac{1}{t}\int_0^t q_\alpha(\theta_s)ds \quad \text{w.p.1} \\ &= \int_{S^1} q_\alpha(\theta)d\mu_\alpha(\theta) \quad \text{w.p.1.} \end{aligned} \tag{5.7}$$

Since (5.7) is valid for all $x \in M$ and $v \in T_xM \setminus \{0\}$ we have

$$\lambda_1(\alpha) = \int_{S^1} q_\alpha(\theta)d\mu_\alpha(\theta). \tag{5.8}$$

Formula (5.8) is a special case of Carverhill's formula. In general the integral is taken over SM, the unit sphere bundle of M. In our case $SM = M \times S^1$ and by the remark after (5.3) and (5.4) it turns out that both the integrand and the density of the measure depend only upon the second factor in the product $M \times S^1$.

When $\alpha = \frac{\pi}{2}$ then $g_\alpha(\theta) = k(\sin^2\alpha + \sin^2(2\theta - \alpha))^{-1/2}$ where k is chosen to ensure total mass 1. Then

$$\begin{aligned}\lambda_1(\tfrac{\pi}{2}) &= -\tfrac{1}{2} + E(2^{-1/2})/K(2^{-1/2}) \\ &= 4\pi^2(\Gamma(\tfrac{1}{4}))^{-4} \\ &\sim 0{\cdot}228\end{aligned}$$

where K and E denote complete elliptic integrals of the first and second kinds (see Whittaker and Watson [26]). For $0 < \alpha < \frac{\pi}{2}$ we do not have such a nice formula for $g_\alpha(\theta)$. Notice that for $0 < \alpha \leqslant \frac{\pi}{2}$ both $q_\alpha(\theta)$ and $g_\alpha(\theta)$, and hence also $\lambda_1(\alpha)$, depend analytically upon α. However as $\alpha \to 0$ the generator L_1 becomes singular and we need to investigate $g_\alpha(\theta)$ in more detail in order to describe the behaviour of $\lambda_1(\alpha)$ as $\alpha \to 0$.

Since the generator L_1 is invariant under the transformation $\theta \mapsto \theta + \frac{\pi}{2}$ it follows that the density g_α has period $\frac{\pi}{2}$. We shall give a formula for $g_\alpha(\theta)$ for $\frac{\alpha}{2} - \frac{\pi}{4} \leqslant \theta \leqslant \frac{\alpha}{2} + \frac{\pi}{4}$. Define

$$f_\alpha(\phi) = (\sin^2\alpha + \sin^2\phi)^{-1/2}\exp\left(\frac{\cos\alpha}{\sqrt{(1+\sin^2\alpha)}} \arctan\left(\frac{\sqrt{(1+\sin^2\alpha)}}{\sin\alpha}\tan\phi\right)\right) \tag{5.9}$$

for $-\frac{\pi}{2} \leqslant \phi \leqslant \frac{\pi}{2}$. Then

$$g_\alpha(\theta) = k_\alpha(\sin^2\alpha + \sin^2(2\theta - \alpha))^{-1}(f_\alpha(2\theta - \alpha))^{-1}\left(1 + \ell_\alpha\int_{-\pi/2}^{2\theta-\alpha} f_\alpha(\phi)\,d\phi\right) \tag{5.10}$$

for $\frac{\alpha}{2} - \frac{\pi}{4} \leqslant \theta \leqslant \frac{\alpha}{2} + \frac{\pi}{4}$, where ℓ_α is chosen so that $g_\alpha(\frac{\alpha}{2} - \frac{\pi}{4}) = g_\alpha(\frac{\alpha}{2} + \frac{\pi}{4})$ and then k_α is chosen to ensure total mass 1. Since

$$\begin{aligned}1 \leqslant 1 + \ell_\alpha\int_{-\pi/2}^{2\theta-\alpha} f_\alpha(\phi)\,d\phi &\leqslant 1 + \ell_\alpha\int_{-\pi/2}^{\pi/2} f_\alpha(\phi)\,d\phi \\ &= \exp\left(\frac{\pi\cos\alpha}{\sqrt{(1+\sin^2\alpha)}}\right) \leqslant e^\pi\end{aligned}$$

we obtain from (5.9), (5.10)

$$\frac{k_\alpha e^{-\pi/2}}{\sqrt{(\sin^2\alpha + \sin^2(2\theta-\alpha))}} \leqslant g_\alpha(\theta) \leqslant \frac{k_\alpha e^{3\pi/2}}{\sqrt{(\sin^2\alpha + \sin^2(2\theta-\alpha))}}.$$

Therefore

$$\frac{1}{2}e^{-2\pi}\frac{E(\gamma)}{\gamma^2 K(\gamma)} \leqslant \lambda_1(\alpha) + \frac{1}{2} \leqslant \frac{1}{2}e^{2\pi}\frac{E(\gamma)}{\gamma^2 K(\gamma)}$$

where $\gamma = (1 + \sin^2\alpha)^{-1/2}$. Now as $\alpha \to 0$ then $\sqrt{1-\gamma^2} \sim \alpha$ and so

$\frac{E(\gamma)}{\gamma^2 K(\gamma)} \sim 1/\log(1/\alpha)$ (see [25], p.521). It follows that $\lambda_1 : [0,\frac{\pi}{2}] \to \mathbb{R}$ is continuous but not differentiable at $\alpha = 0$.

Returning to (5.3) we obtain by Itô's lemma

$$d(\log \det A_t) = (\cos\alpha)dU_t^1 + (\cos\alpha)dU_t^2 - (\cos^2\alpha)dt$$

so that

$$\lambda_1 + \lambda_2 = \lim_{t\to\infty} \frac{1}{t} \log \det A_t = -\cos^2\alpha. \qquad (5.11)$$

Combining (5.6), (5.8) and (5.11) we have

$$\lambda_i(\alpha) = \frac{1}{2}\left(-\cos^2\alpha + (-1)^{i-1}\int_{S^1} \sin^2(2\theta-\alpha)d\mu_\alpha(\theta)\right). \qquad (5.12)$$

The table gives values of $\lambda_i(\alpha)$ for α at intervals of 0·1 radians. Values of $\int_{S^1} \sin^2(2\theta-\alpha)d\mu_\alpha(\theta)$ were obtained by numerical integration and substituted into (5.12). We obtain $\lambda_1(\alpha) < 0$ for

α	$\lambda_1(\alpha)$	$\lambda_2(\alpha)$
0	-0·5	-0·5
0·1	-0·346	-0·644
0·2	-0·307	-0·653
0·3	-0·268	-0·644
0·4	-0·226	-0·622
0·5	-0·179	-0·591
0·6	-0·129	-0·552
0·7	-0·077	-0·508
0·8	-0·024	-0·462
0·9	0·029	-0·415
1·0	0·078	-0·370
1·1	0·123	-0·328
1·2	0·161	-0·292
1·3	0·192	-0·263
1·4	0·214	-0·243
1·5	0·226	-0·231
$\pi/2$	0·228	-0·228

$\alpha < 0{\cdot}845 = 0{\cdot}269\pi$ radians. From (5.11) we have $\lambda_1(\alpha) + \lambda_2(\alpha) < 0$ for $\alpha < \frac{\pi}{2}$, so that $\alpha = \frac{\pi}{2}$ is the only case in which the measure ρ is invariant under the flow. In all other cases $I(\rho_t|\rho) \to \infty$ as $t \to \infty$ w.p.1. In particular the Radon-Nikodym derivative $\frac{d\rho_t}{d\rho}$ (which exists and is a smooth function on M since ξ_t is a diffeomorphism) becomes unbounded as $t \to \infty$ w.p.1. Roughly speaking, ρ_t becomes more and more singular with respect to ρ as $t \to \infty$.

Finally we return to the four special cases which we saw at an early stage in Figures 1 to 4. If $\alpha = 0$ then $\operatorname{diam}\{\xi_t(x) : x \in J\} \to 0$ as $t \to \infty$ w.p.1. The picture of a simulation which has been running

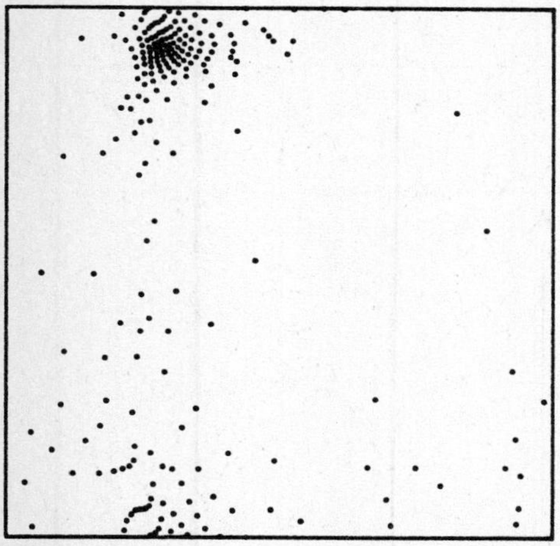

Figure 5 : $\alpha = \pi/6$, $t = 2{\cdot}5$

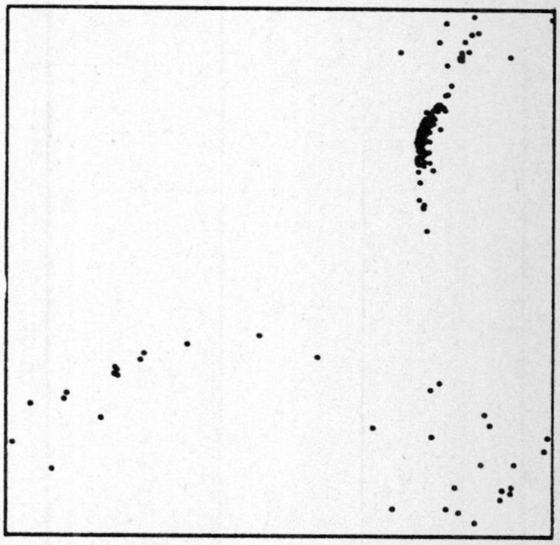

Figure 6 : $\alpha = \pi/6$, $t = 5$

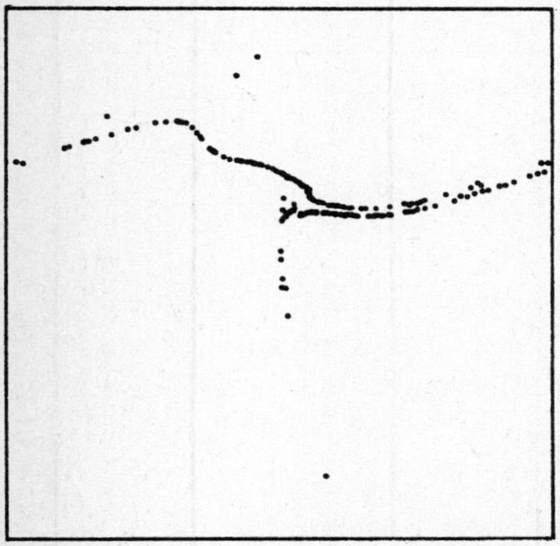

Figure 7 : $\alpha = \pi/6$, $t = 10$

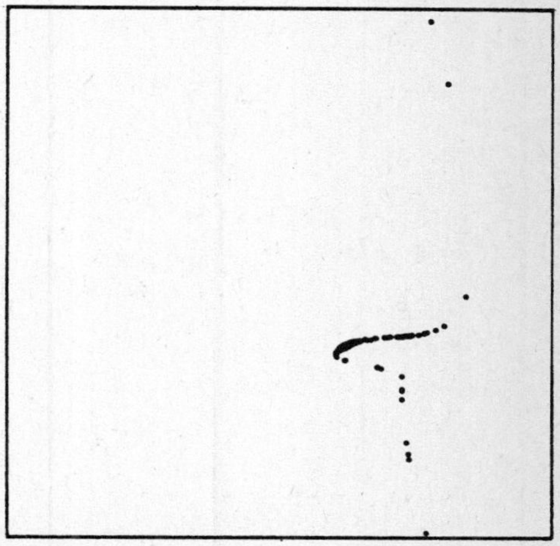

Figure 8 : $\alpha = \pi/6$, $t = 20$

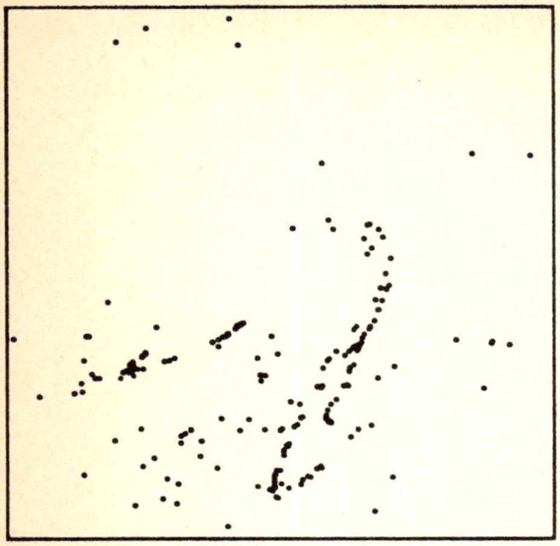

Figure 9 : $\alpha = \pi/3$, $t = 100$

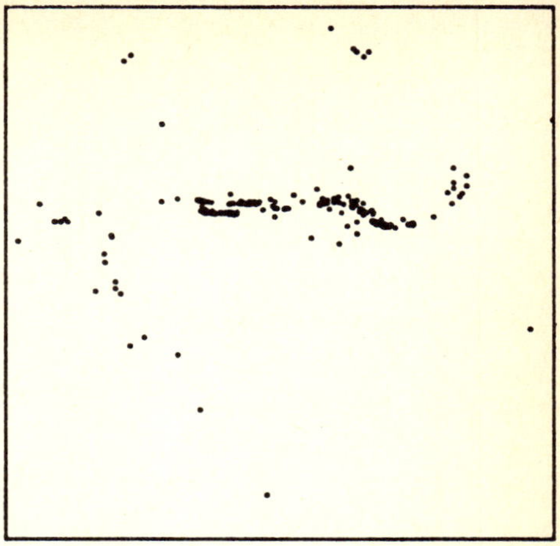

Figure 10 : $\alpha = \pi/3$, $t = 105$

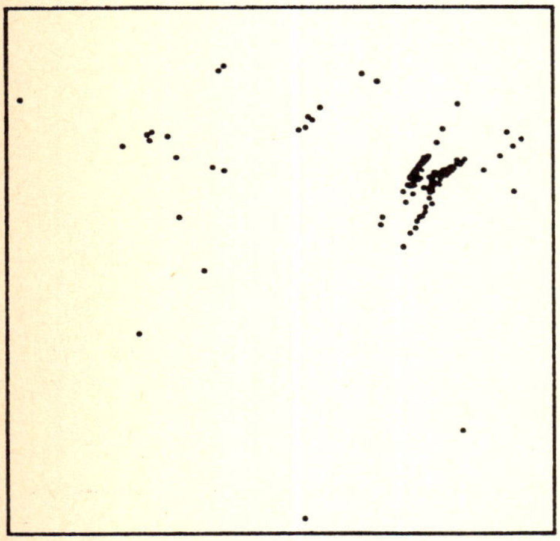

Figure 11 : $\alpha = \pi/3$, $t = 110$

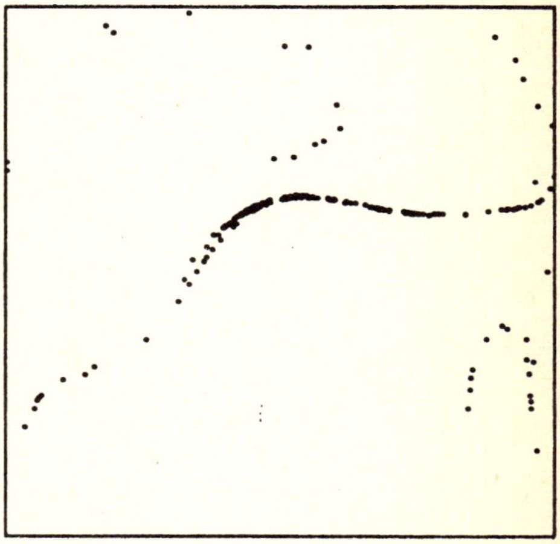

Figure 12 : $\alpha = \pi/3$, $t = 115$

for a long time will show a single point moving in M like a Brownian particle. The single point is really 225 distinct points which are very close together. If $\alpha = \frac{\pi}{2}$ it can be shown that the distribution of the random set $\{\xi_t(x) : x \in J\}$ converges as $t \to \infty$ to that obtained by choosing 225 independent points in M distributed according to ρ. This is because ρ^{225} is the unique stationary probability measure for the 225-point process on $\{(x_1,\dots,x_{225}) \in M^{225} : x_i \neq x_j \text{ whenever } i \neq j\}$.

Figures 5 to 8 and 9 to 12 are taken from simulations for $\alpha = \frac{\pi}{6}$ and $\alpha = \frac{\pi}{3}$ respectively. Although Lyapunov exponents are defined in terms of almost-sure limiting behaviour as $t \to \infty$ it turns out that the simulations (at finite times) give pictures which can be explained in terms of the values of the Lyapunov exponents. When $\alpha = \frac{\pi}{6}$ we have $\lambda_1 = -0{\cdot}168$ and $\lambda_2 = -0{\cdot}583$. The fact that the original lattice appears to be compressed relatively quickly to a string of points which then contracts along its length corresponds to the two different rates of contraction given by $\lambda_2 < \lambda_1 < 0$. Eventually this simulation will show a single Brownian particle on M. When $\alpha = \frac{\pi}{3}$ we have $\lambda_1 = 0{\cdot}100$ and $\lambda_2 = -0{\cdot}350$. Figures 9 to 12 show the simulation at the comparatively large times 100, 105, 110 and 115. The distribution of points is biased towards the existence of small strings of points, but is not converging to a single dense cluster of points. The existence for short intervals of time of clusters of points can be attributed to the fact that $\lambda_1 + \lambda_2 < 0$. The tendency of any cluster of points to be elongated into a string (for example in the transition from Figure 11 to Figure 12) corresponds to the fact that $\lambda_1 > 0$.

Acknowledgement I wish to thank Dr John Pulham of the Department of Mathematics of the University of Aberdeen for providing Figures 1 to 12 and the data for the table of values of λ_1 and λ_2.

References

1. P.H. Baxendale, Wiener processes on manifolds of maps, Proc. Royal Soc. Edinburgh 87A (1980), 127-152.

2. P.H. Baxendale, Brownian motions in the diffeomorphism group I, Compositio Math. 53 (1984), 19-50.

3. P.H. Baxendale, Asymptotic behaviour of stochastic flows of diffeomorphisms: two case studies, preprint, Aberdeen University, 1984.

4. P.H. Baxendale, The Lyapunov spectrum of a stochastic flow of diffeomorphisms, Proceedings of a Workshop on Lyapunov exponents, ed. L. Arnold and V. Wihstutz, Springer Lecture Notes in Mathematics, 1985.

5. P.H. Baxendale and T.E. Harris, Isotropic stochastic flows, Ann. Probab. (to appear).

6. J-M. Bismut, Mécanique aléatoire, Springer Lecture Notes in Mathematics, 1981.

7. A.P. Carverhill, Flows of stochastic dynamical systems: ergodic theory, Stochastics 14 (1985), 273-318.

8. A.P. Carverhill, A formula for the Lyapunov numbers of a stochastic flow. Application to a perturbation theorem, Stochastics 14 (1985), 209-226.

9. A.P. Carverhill, Survey: Lyapunov exponents for stochastic flows on manifolds, Proceedings of a Workshop on Lyapunov exponents, ed. L. Arnold and V. Wihstutz, Springer Lecture Notes in Mathematics, 1985.

10. A.P. Carverhill, M.J. Chappell and K.D. Elworthy, Characteristic exponents for stochastic flows, Proceedings of BIBOS I: Stochastic Processes, Springer Lecture Notes in Mathematics, 1985.

11. M.J. Chappell, Bounds for average Lyapunov exponents of gradient stochastic systems, Proceedings of a Workshop on Lyapunov exponents, ed. L. Arnold and V. Wihstutz, Springer Lecture Notes in Mathematics, 1985.

12. K.D. Elworthy, Stochastic differential equations on manifolds, Cambridge University Press, 1982.

13. H. Furstenberg, Noncommuting random products, Trans. Amer. Math. Soc. 108 (1963), 377-428.

14. R.Z. Has'minskii, Necessary and sufficient conditions for the asymptotic stability of linear stochastic systems, Theory Prob. Appl. 12 (1967), 144-147.

15. N. Ikeda and S. Watanabe, Stochastic differential equations and diffusion processes, North Holland, Amsterdam, 1981.

16. N. Ikeda and S. Watanabe, Stochastic flows of diffeomorphisms, Advances in Probability 7 (Stochastic Analysis and Applications, ed. M. Pinsky) (1984), 179-198.

17. S. Kakutani, On equivalence of infinite product measures, Ann. of Math. 49 (1948), 214-224.

18. S. Kobayashi, Transformation groups in differential geometry, Springer-Verlag, Berlin, 1972.

19. H. Kunita, Stochastic differential equations and stochastic flow of diffeomorphisms, École d'Été de Probabilités de Saint-Flour XII, Springer Lecture Notes in Mathematics, 1984, 143-303.

20. Y. Le Jan, Équilibre et exposants de Lyapunov de certain flots browniens, C.R. Acad. Sc. Paris Sér I 298 (1984), 361-364.

21. Y. Le Jan, Exposants de Lyapunov pour les mouvements browniens isotropes, C.R. Acad. Sc. Paris Sér I 299 (1984), 947-949.

22. Y. Le Jan, Equilibrium state for a turbulent flow of diffusion, Proceedings: Stochastic processes and infinite dimensional analysis, Bielefeld 1983, Pitman Research Notes in Mathematics 124, 1985.

23. Y. Le Jan and S. Watanabe, Stochastic flows of diffeomorphisms, Proceedings of Taniguchi symposium on stochastic analysis, Katata 1982, 307-332.

24. V.I. Oseledec, A multiplicative ergodic theorem. Lyapunov characteristic numbers for dynamical systems, Trans. Moscow Math. Soc. 19 (1968), 197-231.

25. D. Ruelle, Ergodic theory of differentiable dynamical systems, Inst. Hautes Etudes Sci. Publ. Math. 50 (1979), 275-305.

26. E.T. Whittaker and G.N. Watson, A course of modern analysis, Cambridge University Press, 1950.

STOCHASTIC ENSEMBLES AND HIERARCHIES

Donald Dawson

Department of Mathematics & Statistics
Carleton University,
Ottawa, Canada K1S 5B6.

1. INTRODUCTION.

Large scale interacting systems arise in population biology, statistical physics, neural network models, economics, power systems, etc. Generally large scale nonlinear systems are highly intractable and can exhibit extremely complex behavior including multiple equilibria and large scale fluctuations leading from one stable regime to another. In order to gain some understanding of these phenomena model stochastic systems have been devised which can exhibit some of these behaviors but which satisfy certain simplifying assumptions. One such model involves the study of N symmetrically interacting subsystems; the $N \to \infty$ limit (known as the McKean-Vlasov limit) of such a system often leads to a Markov process with nonlinear generator. In this work we modify this idea by building models with multi-level symmetries. This leads to the notion of a hierarchy of interacting systems in which the interaction between the subsystems decreases in order of magnitude as the level in the hierarchy increases. Furthermore at each level of the hierarchy we assume a symmetric structure. The symmetry assumption is warranted provided that the state space for the interacting individuals is sufficiently rich to determine their interaction behavior and is therefore analogous to the considerations involved in building Markov models.

The mathematical formulation given in the first part of the paper is in the setting of hypermeasure-valued processes; for example, a two-level system is described as a $M_2(S) := M_1(M_1(S))$-valued Markov process where $M_1(S)$ denotes the space of probability measures on S. In the $N \to \infty$ limit of a single level ensemble the limit process is a measure-valued diffusion or in the degenerate case a deterministic measure-valued process which represents the probability law of an S-valued process which is described by a nonlinear martingale problem. These systems determine new types of stochastic process which combine the features of nonlinear martingale problems and infinite dimensional Markov processes. We introduce and illustrate a method to establish existence and uniqueness of solutions to certain martingale problems of this type. This method is based on the use of dual processes at two or more levels which make it possible to successively build more complex systems and is thus a type of "bootstrap" method. In order to illustrate these ideas we consider a model suggested by population genetics. This model can serve to illustrate some of the ideas and techniques useful for this family of models. The ultimate goal of this research is to describe the nature of the flow of organization and control between levels in the hierarchy. Useful tools in the study of these questions include scaling in space and time, bifurcation theory, multiple time scales, averaging, singular perturbation and large deviation theory.

2. A MATHEMATICAL FRAMEWORK FOR HYPERMEASURE-VALUED PROCESSES.

Let (S,d) be a compact metric space, $M_1(S)$ denote the space of probability measures on S and $C(S)$ denote the space of continuous functions on S. Let $L_1(S)$ denote the subspace of $C(S)$ consisting of functions satisfying

$$\|f\|_1 := \sup_{x \neq y} (|f(x) - f(y)| / d(x,y)) \leq 1.$$

The weak topology on $M_1(S)$ is defined by $\mu_n \xrightarrow{w} \mu$ as $n \to \infty$ if and only if for every $f \in C(S)$, $\lim_{n\to\infty} \langle f,\mu_n\rangle = \langle f,\mu\rangle$, where $\langle f,\mu\rangle := \int f(x)\mu(dx)$. $M_1(S)$ can be furnished with several metrics compatible with this topology. Two examples are:

(i) the dual Lipschitz metric which is defined by

$$\rho_L(\mu_1,\mu_2) := \sup_{f \in L_1(S)} \left(\int_S f d\mu_1 - \int_S f d\mu_2\right);$$

(ii) $\rho_V(\mu_1,\mu_2) := \inf \int_S \int_S d(x,y) m(dx,dy)$

where the infimum is taken over all measures m on $S\times S$ with marginals μ_1 and μ_2.

It turns out that these two metrics are actually identical, that is, $\rho = \rho_L = \rho_V$ (cf. Zolotarev (1984)).

Note that $(M_1(S),\rho)$ is a compact metric space and therefore we can define recursively the hierarchy:

$M_1 := M_1(S)$,

$M_{k+1} := M_1(M_k)$ for $k \geq 1$

and the metric ρ_{k+1} on M_{k+1} is defined as above. The elements of M_k for $k > 1$ are called <u>k-level hypermeasures</u>. Given $f \in C(S)$ we denote by $f^{(k)}$ the function defined on M_k recursively as follows:

$f^{(0)} := f$, $f^{(1)}(\mu) := \langle f,\mu\rangle$, and for $k \geq 1$, $\mu \in M_k$,

$f^{(k+1)}(\mu) := \langle f^{(k)},\mu\rangle$.

The metric ρ_{k+1} can be characterized as follows.

<u>LEMMA 2.1</u>. For $\mu_1, \mu_2 \in M_{k+1}$,

(2.1) $\rho_{k+1}(\mu_1,\mu_2) = \inf \langle d^{(k)}, m(\mu_1,\mu_2)\rangle$

where the infimum is over all measures in $M_{k+1}(S\times S)$ with marginals μ_1 and μ_2 and the metric $d(.,.)$ is regarded as a function on $S\times S$.

Proof. See Dawson (1985).

A sequence $\mu_n \in M_{k+1}$ converges weakly to $\mu \in M_{k+1}$, $\mu_n \xrightarrow{w} \mu$ if for every $f \in C(M_k)$,

(2.2) $\lim_{n\to\infty} \langle f,\mu_n\rangle = \langle f,\mu\rangle$.

It is sufficient to verify (2.2) for $f \in D_k$ provided that D_k is a dense subset of $C(M_k)$. Since M_k is compact, the Stone Weierstrass theorem implies that an algebra of functions that separates points is dense in $C(M_k)$. Let D_0 denote a dense subset of $C(S)$.

<u>LEMMA 2.2.</u> The smallest algebras of functions containing each of the following classes is dense in $C(M_{k+1})$.

(a) $F(\mu) = f(\langle\phi,\mu\rangle)$ with $f \in C(R^1)$, $\phi \in D_k$, where D_k is a dense subset of $C(M_k)$,

(b) $F(\mu) = \langle h_1,\mu\rangle\langle h_2,\mu\rangle \ldots \langle h_n,\mu\rangle$, $n \in Z^+$, $h_i \in D_k$ for $i = 1,\ldots,n$.

Proof. It suffices to show that each of these classes of functions separates points. But this follows since if $\langle\phi,\mu_1\rangle = \langle\phi,\mu_2\rangle$ for all $\phi \in D_k$, then $\mu_1 = \mu_2$.

<u>REMARK 2.1</u>. In the case of functions of the form (b) it suffices to take linear combinations since these classes are closed under multiplication.

<u>COROLLARY</u>. The smallest algebra of functions containing those in either of the following two classes is dense in $C(M_2)$:

(c) $F(\nu) = f(\langle h_2(\langle h_1,.\rangle),\nu\rangle)$ with $h_1 \in D_0$, $h_2, f \in C_b(R^1)$, the class of bounded continuous functions on R^1 and $\nu \in M_2$;

(d) $F(\nu) = \langle(\langle h_{1,1},\mu\rangle\ldots\langle h_{1,n_1},\mu\rangle),\nu\rangle\ldots\langle(\langle h_{m,1},\mu\rangle\ldots\langle h_{m,n_m},\mu\rangle),\nu\rangle$.

Proof. Let D_1 denote the class of functions on M_1 of the form $h_2(\langle h_1,\mu\rangle)$ where $h_1 \in D_0$, $h_2 \in C_b(R^1)$. Then by Lemma 2.2 these functions are dense in $C(M_1)$. Applying Lemma 2.2 a second time we conclude that functions of the form (c) separate points of M_2.

To verify that functions of the form (d) separate points note that linear combinations of functions of the form $(\langle h_{1,1},\mu\rangle, \ldots, \langle h_{1,n_1},\mu\rangle)$ are dense in $C(M_1)$. This completes the proof.

REMARK 2.2. The following class which contains all functions of the form (d) are used in the applications below.

$$F(\nu) = \int \ldots \int \prod_{j=1}^{m} \prod_{i=1}^{n_j} h(x_{1,1},\ldots,x_{1,n_1};x_{2,1},\ldots,x_{2,n_2};\ldots;x_{m,1},\ldots,x_{m,n_m}) \cdot \mu_j(dx_{j,i})\nu(d\mu_j) \quad .$$

The canonical sample spaces for the study of hypermeasure-valued processes are $\Omega_k^D := D(R^+,M_k)$, the space of cadlag functions from R^+ to M_k furnished with the Skorohod topology and $\Omega_k^C := C(R^+,M_k)$, the space of continuous functions from R^+ to M_k furnished with the topology of uniform convergence on compact subsets of R^+. For $\omega \in \Omega_k^D$ let $X(t,\omega) := \omega(t) \in M_k$.

In proving the relative weak compactness of a family of M_k-valued processes it is convenient to make use of the following lemma which reduces the problem to verifying that a certain class of real valued processes are relatively weakly compact.

LEMMA 2.3. Let P_n, $n \in Z^+$, be a sequence of probability measures on Ω_k^D and let D_{k-1} be a dense subset of $C(M_{k-1})$. In order that $\{P_n\}$ be relatively weakly compact in $M_1(\Omega_k^D)$ it suffices that $\{P_n^f\}$, where P_n^f is the law of $\langle f,X(t)\rangle$ under P_n, be uniformly tight in $M_1(\Omega_0^D)$ for each $f \in D_{k-1}$.

Proof. Refer to Dawson (1985).

REMARK 2.3. THE INFINITE LEVEL HIERARCHY.

Note that the expectation operator is a mapping from M_{k+1} onto M_k. Given a probability measure P on M_{k+1},

$$M_1 \xleftarrow{E} M_2 \xleftarrow{E} M_3 \ \ldots \ M_k \xleftarrow{E} M_{k+1}$$

where if $\nu \in M_{k+1}$,

$$E(\nu) = \int_{M_k} \mu\nu(d\mu) \in M_k.$$

Given $\nu \in M_{k+1}$ the sequence can also be continued upwards as follows:

$\nu \longrightarrow \delta_\nu \in M_{k+1}$, where δ_ν denotes the unit mass on the point ν. Therefore any μ in the disjoint union $\bigcup_{k=1}^{\infty} M_k$ can also be interpreted as an element of the projective limit. The projective limit, M_∞, consists of all sequences $\{\mu_k : \mu_k \in M_k, k \geq 1\}$ such that $\mu_k = E\mu_{k+1}$ for each $k \geq 1$. M_∞ is furnished with the coarsest topology such that

$$\{\mu_k^{(n)}\} \longrightarrow \{\mu_k\} \quad \text{in } M_\infty \quad \text{as} \quad n \to \infty,$$

if for every $k \geq 1$, $\mu_k^{(n)} \xrightarrow{w} \mu_k$ as $n \to \infty$.

Examples of infinite level systems are Dyson's hierarchical model in statistical physics (see Sinai (1982)) and the hierarchically clustered population model of Sawyer and Felsenstein (1982).

In this paper hypermeasure-valued processes are characterized as solutions of well-posed martingale problems. In particular a hypermeasure-valued diffusion process is associated with a martingale problem specified by a second order differential operator. In order to define a large class of differential operators we begin with the definition of the first and second derivatives. Let F denote a function defined on M_k. The first derivative is defined by:

$$\delta F(\mu)/\delta\mu(\eta) \quad := d/d\varepsilon(\ F(\mu + \varepsilon\delta_\eta))|_{\varepsilon=0} \quad \text{for } \eta \in M_{k-1},\ \mu \in M_k,$$

provided that that the limit exists and belongs to $C(M_{k-1})$. Similarly the second derivative is defined as a function on $M_{k-1} \times M_{k-1}$ as follows:

$$\delta^2 F(\mu)/\delta\mu(\eta_1)\delta\mu(\eta_2) \ := \partial^2/\partial\varepsilon_1\partial\varepsilon_2\ F(\mu + \ \varepsilon_1\delta_{\eta_1} + \varepsilon_2\delta_{\eta_2})|_{\varepsilon_1=\varepsilon_2=0} \ ,$$

provided that the lätter exists and belongs to $C(M_{k-1} \times M_{k-1})$. Functions of this type are said to belong to $C^2(M_k)$.

The basic results of Stroock and Varadhan (1979) can be applied to the construction of measure-valued processes (Dawson (1985)). In order to construct a M_{k+1}-valued diffusion it suffices to verify that the martingale problem posed by the pair $(C^2(M_{k+1}),\ G_{k+1})$ where G_{k+1} is a second order differential operator acting on $C^2(M_{k+1})$ is well-posed, that is, it has a unique solution consistent with each initial measure. The general form of the generator G_{k+1} is given by:

$$(2.3)\quad G_{k+1}F(\nu) = \int G_k(\nu)\delta F(\nu)/\delta\nu(\mu)\ \nu(d\mu)$$
$$+ \iint B\times B\ (\delta^2 F(\nu)/\delta\nu(\mu_1)\delta\nu(\mu_2))Q_k(\nu{:}d\mu_1 \times d\mu_2)\ ,$$

that is, the sum of a non-constant coefficient first order (drift) operator and a non-constant coefficient second order (diffusion) operator. In turn,

$$(2.4)\quad G_k(\nu)f(\mu) = \int G_{k-1}^{\nu}(\mu)\ \delta f(\mu)/\delta\mu(\eta)\ \mu(d\eta)$$
$$+ \iint b\times b(\delta^2 f(\mu)/\delta\mu(\eta_1)\delta\mu(\eta_2))Q_{k-1}^{\nu}(\mu{:}d\eta_1 \times d\eta_2)\ .$$

In (2.3) B is a operator acting on $C(M_k)$ (or a subspace of $C(M_k)$) and Q_k is a continuous mapping from M_{k+1} to measures on $M_k \times M_k$, and $G_k(\nu)$ is an operator acting on a subspace of $C(M_k)$. In (2.4) and each $\nu \in M_{k+1}$ b, Q_{k-1}^{ν} and G_{k-1}^{ν} are the analogous objects but one level lower in the hierarchy.

REMARK 2.4. If G_{k-1}^{ν} and Q_{k-1}^{ν} in (2.4) do not depend on ν, then G_k is the generator of a M_k-valued diffusion. On the other hand if $G_k(\nu)$ does depend on ν, then it can be interpreted as the nonlinear generator of a McKean-Vlasov type nonlinear Markov process to be defined below.

DEFINITION 2.1. The family $\{P_\mu : \mu \in M_{k+1}\}$ is a solution to the (linear) martingale problem associated with G_{k+1} if for $\mu \in M_{k+1}$, P_μ is a probability measure on $D([0,\infty), M_{k+1})$ with $P_\mu(X(0) = \mu) = 1$ and for which for $f \in C^2(M_{k+1})$,

$$(2.5)\quad f(X(t)) - \int_0^t G_{k+1} f(X(s))ds \text{ is a } P_\mu\text{-martingale for each } \mu \in M_{k+1}.$$

DEFINITION 2.2. The family $\{P_\mu : \mu \in M_{k+1}\}$ is a solution to the non-linear martingale problem associated with G_k^{ν} if P_μ is a probability measure on $D([0,\infty), M_k)$ and

(i) μ is the probability law of $X(0) \in M_k$, and

(ii) $f(X(t)) - \int_0^t G_k^{\Pi_s P_\mu} f(X(s))ds$ is a P_μ-martingale

where $\Pi_s : M_1(D([0,\infty), M_k)) \to M_1(M_k)$ by $\Pi_s P(B) := P(\{\omega : X(s,\omega) \in B\})$.

REMARK 2.5. Note that in the case of a nonlinear martingale problem we must specify the solution measure for all $\mu \in M_{k+1}$, that is, all initial distributions and not simply all initial points. The reason for this is that in the case of a solution to a nonlinear martingale problem the linear property $P_\mu = \int P_x \mu(dx)$ need not be satisfied.

3. EXAMPLES OF STOCHASTIC ENSEMBLES AND HIERARCHIES.

EXAMPLE 3.1. SINGLE LEVEL NON-INTERACTING DIFFUSIONS.

For $i = 1,2,\ldots,N$ let $x_i(.)$ satisfy the stochastic differential equations

$$(3.1)\quad dx_i(t) = b(x_i(t))dt + \sigma(x_i(t))dw_i(t)$$

where $\{w_i(.):i = 1,...,N\}$ are independent R^d-valued Brownian motions. The coefficients $b(.)$ and $\sigma(.)$ are assumed to satisfy the usual Lipschitz and growth conditions to assure the existence of a unique strong solution. The <u>empirical distribution process</u> is defined by

$$(3.1) \qquad X_N(t) := N^{-1} \sum_{i=1}^{N} \delta_{x_i(t)} .$$

$X_N(.)$ is a $M_1(R^d)$-valued Markov process; for convenience it is useful to consider it as a process in $M_1(S)$ where S is the one point compactification of R^d. To prescribe the generator of X_N we consider test functions of the form $F(\mu) = f(<\phi,\mu>)$ where $<\phi,\mu> = \int\phi(x)\mu(dx)$. The generator for X_N is then given by:

$$(3.2) \qquad G_N F(\mu) = G_1 F(\mu) + N^{-1} G_2 F(\mu), \quad \text{where}$$

$$G_1 F(\mu) = \int(L(\delta F(\mu)/\delta\mu(x))\mu(dx) = f'(<\phi,\mu>)<L\phi,\mu> ,$$

$$G_2 F(\mu) = \tfrac{1}{2} f''(<\phi,\mu\ )<|\nabla\phi|_q^2,\mu> ,$$

$$L\phi(x) = b(x).\nabla\phi(x) + \tfrac{1}{2} \sum_{i,j=1}^{d} q_{ij}(x)\partial^2\phi/\partial x_i\partial x_j ,$$

$$|\nabla\phi|_q^2 = \sum_{i,j=1}^{d} q_{ij}\ \partial\phi/\partial x_i\ \partial\phi/\partial x_j , \text{ and}$$

$q := \sigma\sigma^T$ where σ^T is the transpose of σ and $.$ denotes the scalar product in R^d.

<u>EXAMPLE 3.2</u>. SINGLE LEVEL SYMMETRIC INTERACTIONS.

In this example the previous one is modified by incorporating symmetric interactions - this is achieved by allowing the drift and diffusion coefficients b and σ to depend on the empirical distribution. In particular let

$$b: R^d \times M_1(S) \to R, \qquad \sigma: R^d \times M_1(S) \to R^+.$$

Then $$L(\mu)\phi = b(x,\mu).\nabla\phi(x) + \tfrac{1}{2} \sum_{i,j=1}^{d} q_{ij}(x,\mu)\partial^2\phi/\partial x_i\partial x_j ,$$

where $q(\mu) := \sigma(\mu)\sigma(\mu)^T$. Then

$$(3.3) \qquad G_N F(\mu) = G_1 F(\mu) + N^{-1} G_2 F(\mu)$$

$$= f'(<\phi,\mu>)<L(\mu)\phi,\mu> + \tfrac{1}{2}N^{-1} f''(<\phi,\mu>)<|\nabla\phi|_{q(\mu)}^2,\mu> .$$

<u>EXAMPLE 3.3</u>. ADDING A REPLACEMENT INTERACTION.

In this version individuals can be removed and replaced by new individuals by random sampling from the empirical distribution. In this case

$$(3.4) \qquad G_N F(\mu) = G_1 F(\mu) + N^{-1} G_2 F(\mu) + G_3 F(\mu),$$

where

$$G_3F(\mu) = \tfrac{1}{2}\iint[F(\mu - N^{-1}\delta_x + N^{-1}\delta_y) - F(\mu)]\Gamma_N(\mu;y)\mu(dx)\mu(dy)$$

where $\Gamma_N(\mu;y) = N^2\Gamma + N\Gamma^*(\mu;y)$ and $\Gamma^*(\mu;y)$ denotes the "selective advantage" of an individual of type y. The operator corresponds to a system with a "diffusion limit scaling". Then

$$(3.5) \quad G_3F(\mu) = \tfrac{1}{2}\Gamma\iint[\delta^2F(\mu)/\delta\mu(x)^2 + \delta^2F(\mu)/\delta\mu(y)^2 - 2\delta^2F(\mu)/\delta\mu(x)\delta\mu(y)] \cdot\mu(dx)\mu(dy)$$

$$+ \int[\delta F(\mu)/\delta\mu(y)\Gamma^*(\mu;y)]\mu(dy)$$

$$- [\int\delta F(\mu)/\delta\mu(x)\ \mu(dx)].[\int\Gamma^*(\mu;y)\mu(dy)] + R(F)/N,$$

where $R(F)$ denotes a remainder term.

APPLICATION TO A SEARCH ALGORITHM.

This model is a variant of an optimization algorithm due to Ermatov and Zhiglyavskii (1983). Let $V(x)$ be a non-negative function on R^d which has a unique global maximum. A team of N searchers looks for the maximum by following the dynamics:

(i) each searcher moves up the gradient of V,

(ii) each searcher performs a local random search with motility σ_i,

(iii) searchers are removed and replaced by new searchers - the probability that a replacement searcher is of type x_i is $V(x_i)/\sum V(x_i)$.

The state space for this process is $R^d \times R^+$ where the second component prescribes the motilities σ_i. The generator acting on test functions of the form $F(\mu) = f(\langle\phi,\mu\rangle)$ is given by:

$$(3.6) \quad G_NF(\mu) = f'(\langle\phi,\mu\rangle)\langle(\nabla_x V.\nabla_x\phi) + \tfrac{1}{2}\sigma^2\Delta_x\phi,\mu\rangle$$

$$+ f'(\langle\phi,\mu\rangle)[\langle V\phi,\mu\rangle - \langle\phi,\mu\rangle\langle V,\mu\rangle]$$

$$+ \gamma f''(\langle\phi,\mu\rangle)[\langle\phi^2,\mu\rangle - \langle\phi,\mu\rangle^2]$$

$$+ f'(\langle\phi,\mu\rangle)\langle A_t\ \partial\phi/\partial\sigma,\mu\rangle + R(f)/N$$

where the terms describe gradient and diffusion drift, weighted replacement, sampling noise, possibly adaptive change in the distribution of motilities and remainder term, respectively.

EXAMPLE 3.4. A TWO LEVEL HIERARCHY.

Consider a system of M colonies each consisting of N individuals. Let $x_{ij}(t)$ denote the state of the jth individual in the ith colony at time t.

For $i = 1,\dots,M$, let

$$\overline{X}_i(t) := N^{-1}\sum_{k=1}^{N}\delta_{x_{ik}(t)} \in M_1(S).$$

Let

$$(3.7)\qquad \overline{\overline{X}}_{N,M}(t) := M^{-1}\sum_{i=1}^{M}\delta_{\overline{X}_i(t)} \in M_2 = M_1(M_1(S)).$$

Example 3.4.a. For $i = 1,\dots,M$; $j = 1,\dots,N$ let

$$(3.8)\qquad dx_{ij}(t) = [\int b_1(x_{ij}(t),y)\overline{X}_i(t,dy)]dt + b_0(x_{ij}(t))dt$$
$$+ [\int\sigma_1(x_{ij}(t),y)\overline{X}_i(t,dy)]dw_{ij}(t)$$
$$+ \int[\int b_2^{N,M}(x,\eta)\overline{X}_i(t,dx)]\overline{\overline{X}}_{N,M}(t,d\eta)dt$$
$$+ \int[\int\sigma_2(x,\eta)\overline{X}_i(t,dx)]\overline{\overline{X}}_{N,M}(t,d\eta)\; dw_i(t)$$

where $\{w_{ij}(.) : i = 1,\dots,M;\ j = 1,\dots,N\}$ and $\{w_i(.) : i = 1,\dots,M\}$ are independent Brownian motions in R^d. Then $\overline{\overline{X}}_{N,M}(.)$ is an M_2-valued Markov process. To prescribe the generator consider test functions of the form

$$F(\nu) = f(\int_{M_1} h_1[\int_S h_2(x)\mu(dx)]\nu(d\mu)) = f(\langle\langle h_1(\langle h_2,.\rangle),\nu\rangle\rangle)$$

where $\langle\langle h,\nu\rangle\rangle := \int_{M_1} h(\mu)\nu(d\mu)$. For simplicity of exposition we assume that σ_1 and σ_2 are constants. The the generator of the process $\overline{\overline{X}}_{N,M}$ is given by:

$$(3.9)\qquad G_{N,M}F(\nu) = G_1F(\nu) + G_2F(\nu) + G_3F(\nu)$$

where

$$G_1F(\nu) = f'(\nu)\langle\langle h_1'(\mu)\langle\beta_{N,M}(.,\mu,\nu).\nabla h_2,\mu\rangle,\nu\rangle\rangle \text{ where}$$

$$\beta(x,\mu,\nu) := b_0(x) + \int b_1(x,y)\mu(dy) + \int(\int b_2^{N,M}(x,\eta)\mu(dx))\nu(dy)$$

$$G_2F(\nu) = \textit{inside colony fluctuations}$$
$$= \tfrac{1}{2}\,\sigma_1^2 f'(\nu)\langle\langle h_1'(\mu)\langle\Delta h_2,\mu\rangle,\nu\rangle\rangle$$
$$+(\sigma_1^2/2N)f'(\nu)\langle\langle h_1''(\mu)\langle|\nabla h_2|^2,\mu\rangle,\nu\rangle\rangle$$
$$+ (\sigma_1^2/2MN)f''(\nu)\langle\langle h_1'(\mu)\langle|\nabla h_2|^2,\mu\rangle,\nu\rangle\rangle\ ,$$

$$G_3F(\nu) = \textit{between colony fluctuations}$$
$$= (\sigma_2^2/M)f''(\nu)\langle\langle h_1'(\mu)\langle h_2,\mu\rangle^2,\nu\rangle\rangle +$$

$$+ \tfrac{1}{2}\, \sigma_2^2 f'(\nu) \langle\langle h_1''(\mu)\, |\langle \nabla h_2, \mu\rangle|^2, \nu\rangle\rangle$$

$$+ \tfrac{1}{2}\, \sigma_2^2 f'(\nu) \langle\langle h_1'(\mu) \langle \Delta h_2, \mu\rangle, \nu\rangle\rangle \; .$$

Example 3.4.b. Addition of Multilevel replacement of Individuals

In this version we add multilevel replacement of individuals. This leads to the generator:

$$(3.10) \quad G_{N,M}F(\nu) = G_1 + G_2 + G_3 + G_4 + G_5$$

where G_1, G_2 and G_3 are as above and

$$G_4 F(\nu) = \textit{replacement of individuals}$$

$$= \iiint [F(\nu + M^{-1}\{\delta_{\mu - \delta_x/N + \delta_y/N}\}) - F(\nu)]\Gamma_1^N(\mu,y)\mu(dx)\mu(dy) \cdot \nu(d\mu)$$

$$G_5 F(\nu) = \textit{replacement of colonies}$$

$$= \iint [F(\nu + \delta_{\mu_1}/M - \delta_{\mu_2}/M) - F(\nu)]\Gamma_2^{N,M}(\nu,\mu_1)\nu(d\mu_1)\nu(d\mu_2).$$

EXAMPLE 3.5. TWO LEVEL BISTABLE MODEL WITH MEAN-FIELD INTERACTION.

For $i = 1,\ldots,M$; $j = 1,\ldots,N$ let (cf. Dawson 1983)

$$(3.11) \quad dx_{ij}(t) = (-x_{ij}^3(t) + x_{ij}(t))dt + \theta_1 \left(\int x \bar{X}_i(t,dx) - x_{ij}(t)\right)dt$$
$$+ (\theta_2/N^\alpha)\left[\left(\int \left[\int x\mu(t,dx)\right]\bar{\bar{X}}_{N,M}(t,d\mu)\right) - x_{ij}(t)\right]dt$$
$$+ \sigma_1 dw_{ij}(t) + \sigma_2 dw_i(t)$$

where the θ_1 term describes the mean-field interaction, the θ_2 term describes the weak interaction due to feedback between colonies, the σ_1 term denotes the noise perturbation at the level of individuals and the σ_2 term denotes the noise perturbation at the level of colonies. In this case

$$\beta_{N,M}(x,\mu,\nu) = x - x^3 + \theta_1(\langle x,\mu\rangle - x) + (\theta_2/N^\alpha)(\langle\langle(\langle x,\mu\rangle),\nu\rangle\rangle - \langle x,\mu\rangle).$$

REMARK 3.1 HEURISTIC CARICATURE OF DISTRIBUTED SYSTEMS IN R^d.

The term involving θ_2/N^α is inspired by comparison with lattice systems - we give a heuristic argument to motivate this and to relate the value of α to the dimension d. Consider a lattice system with nearest neighbour interaction given by the system: for $i \in Z^d$,

$$(3.12) \quad dx_i(t) = (-x_i^3(t) + x_i(t))dt + \sigma dw_i(t) + \theta \sum_{|i-j|=1} (x_j(t) - x_i(t))dt.$$

The term coupling different sites involves the discrete Laplacian and can be interpreted as a feedback coming from other sites via spatial diffusion. Now consider the total flow into the subsystem whose sites lie in a large cube in R^d of radius r. The flow across the boundary is governed by the smallest eigenvalue of $-\Delta$ with Dirichlet boundary conditions on the boundary of the cube. But in dimension d the smallest eigenvalue is given by $\lambda_1 = c/r^2 = c/V^{2/d}$ where c is a constant and V is the volume of the cube. Thus this suggests that the spatial system can be caricatured by the system (3.11) with $\alpha = 2/d$.

EXAMPLE 3.6. A HIERARCHICALLY CLUSTERED POPULATION MODEL.

In this example we consider a population model involving colonies of N individuals grouped in M colonies and incorporating mutation, sampling of both colonies and individuals and migration between colonies. The generator for this system with diffusion scaling is:

(3.13) $G_{N,M}F(\nu) = G_1F(\nu) + N^2G_2F(\nu) + M^2G_3F(\nu) + MNG_4F(\nu)$, where

$G_1F(\nu)$ = *mutation*

$= f'(\nu)\langle\langle h_1'(\nu)\langle Ah_2,\cdot\rangle,\nu\rangle\rangle$ if $F(\nu) = f(\langle\langle h_1(\langle h_2,.\rangle),\nu\rangle\rangle)$,

and A is assumed to be the infinitesimal generator of a strongly continuous semigroup on $C(S)$,

$G_2F(\nu)$ = *sampling of individuals inside colonies*

$= \gamma_1\iiint[F(\nu+M^{-1}\{\delta_{\mu-\delta_x/N+\delta_y/N}\}) - F(\nu)]\mu(dx)\mu(dy)\nu(d\mu)$,

$G_3F(\nu)$ = *sampling replacement of colonies*

$= \gamma_2\iint[F(\nu+\delta_{\mu_1}/M - \delta_{\mu_2}/M) - F(\nu)]\nu(d\mu_1)\nu(d\mu_2)$,

$G_4F(\nu)$ = *migration of individuals between colonies*

$= \alpha\iiiint[F(\nu+M^{-1}\{\delta_{\mu-\delta_x/N+\delta_y/N}\}) - F(\nu)]\mu(dx)\nu(d\mu)\mu'(dy)\nu(d\mu')$.

4. UNIQUENESS OF SOLUTIONS TO MULTILEVEL MARTINGALE PROBLEMS

According to a basic result of Stroock and Varadhan (1979) a family $\{P_\mu\}$ which is the solution to a well-posed martingale problem corresponds to a Markov process whose generator is an extension of the (pre)generator which defines the martingale problem. Using this approach we obtain Markov processes describing the $N \to \infty$ and $M \to \infty$ limits of some of the examples described in Section 3. Existence of solution follows from the tightness of the system of processes $\bar{\bar{X}}_{N,M}$ - this step is carried out using fairly standard arguments and is not discussed in this paper. In this type of problem the question of uniqueness is usually more difficult and this is the main question to be discussed in this section.

Two ideas are used in establishing the uniqueness - these are duality and bootstrapping.

4.1. THE METHOD OF DUALITY.

In this section we consider a special case of duality which will be used in this paper. For a more detailed discussion of this method in the context of measure-valued processes refer to Dawson and Kurtz (1982) and Dawson (1985).

Let E_1 and E_2 denote two metric spaces, a function(bounded,cont.) $f: E_1 \times E_2 \to R^1$, and a function $g:[0,\infty)\times E_1 \times E_2 \to R^1$. Define two operator-valued functions G_s and $G_s^{\#}$ defined by:

(4.1) $G_s f(x,y) = g_s(x,y)$ (an operator on a subspace of $C(E_1)$)

$G_s^{\#} f(x,y) = g_s(x,y)$ (an operator on a subspace of $C(E_2)$).

THEOREM 4.1. Let $X(.)$, $Y(.)$ be measurable processes with state spaces E_1 and E_2, respectively and assume that $g(.,.)$ is also a bounded continuous function. Assume that

(4.2.1) $f(X(t),y) - \int_0^t G_s f(X(s),y)ds$ is a P_x^X-martingale for every $x \in E_1$ and $y \in E_2$,

(4.2.2) $f(x,Y(t)) - \int_0^t G_s^{\#} f(x,Y(s))ds$ is a P_y^Y-martingale for every $x \in E_1$ and $y \in E_2$.

Then $E_x^X(f(X(t),y)) = E_y^Y(f(x,Y(t))$ for $t \geq 0$ and $x \in E_1$, $y \in E_2$.

Proof. Let $\Phi(s,t) := E\, f(X(s),Y(t))$ where the expectation is with respect to $P = P_x^X \times P_y^Y$ (product measure). Then using a lemma in Dawson and Kurtz (1982),

$$\Phi(t,0) - \Phi(\theta,t) = \int_0^t (\Phi_1(s,t-s) - \Phi_2(s,t-s))ds$$

where $\Phi_1(s,t) = E[g_s(X(s),Y(t))]$, $\Phi_2(t,s) = E[g_s(X(t),Y(s))]$.

Therefore

$$\int_0^t [\Phi_1(s,t-s) - \Phi_2(s,t-s)]ds$$
$$= \int_0^t E[g_s(X(s),Y(t-s))]ds - \int_0^t E[g_{t-s}(X(t-s),Y(s))]ds$$
$$= \int_0^t E[g_s(X(s),Y(t-s)]ds - \int_0^t E[g_{s'}(X(s'),Y(t-s'))]ds' = 0,$$

and the proof is complete.

Let G be a pregenerator of an $M(S)$-valued process and $\{F_h : h \in H\}$ be a dense algebra of functions in $C_b(M(S))$. Let $F(\mu,h) = F_\mu(h) = F_h(\mu)$. Assume that $n : H \to Z^+$ and

$$(4.3) \quad GF_h(\mu) = G^{\#}F_\mu(h) = c_1 \sum_{j=1}^{n(h)} (F_\mu(\Phi_j h) - F_\mu(h)) + c_2 \sum_{j=1}^{n(h)} \sum_{k=1}^{n(h)} (F_\mu(\Phi_{jk}h) - F_\mu(h)) + c_3 F_\mu(Ah)$$

where $\Phi_j : H \to H$, $\Phi_{jk} : H \to H$ with $n(\Phi_j h) \leq n(h)$, $n(\Phi_{jk}h) \leq n(h)$, and and A generates a semigroup of operators on H, $\{T_t\}$, with $n(T_t h) \leq n(h)$ for all $t \geq 0$.

Then it is easy to verify that $G^{\#}$ is the generator of an H-valued Markov process $Y(.)$ with dynamics:

(i) jumps $h \to \Phi_j h$ at rate c_1 for $j = 1,\ldots,n(h)$

(ii) jumps $h \to \Phi_{jk} h$ at rate c_2 for $j,k = 1,\ldots,n(h)$,

(iii) bewteen jumps $Y(.)$ evolves deterministically according to the semigroup T_t.

Let $\{P_\mu : \mu \in M(S)\}$ be a family of measures on $D([0,\infty),M(S))$ which is a solution of the martingale problem associated with the operator G. Then applying Theorem 4.1 it follows that the duality relation

$$(4.4) \quad E_\mu(F_h(X(t)) = E_h(F_\mu(Y(t)) .$$

The relation implies that the solution to the G-martingale problem is unique provided that the family $\{f(.,y) : y \in E_2\}$ is dense in $C_b(E_1)$.

4.2. BOOTSTRAP ARGUMENTS FOR MULTILEVEL MARTINGALE PROBLEMS.

The term bootstrap argument refers to any method that reduces the problem of uniquness to a martingale problem to that of the uniqueness of one or more simpler martingale problems.

To illustrate the ideas involved consider the martingale problem associated with the operator G_{k+1} defined by (2.3). We first consider the simpler problem obtained by setting $Q_k = 0$. Then the resulting degenerate (k+1)-level problem corresponds to a k-level nonlinear martingale problem. Note that a solution to the nonlinear martingale problem induces a solution to the degenerate M_{k+1}-valued martingale problem and the one dimensional distributions of the nonlinear martingale problem are determined by the latter. In fact we can reduce the problem of proving uniqueness to the M_k-valued nonlinear martingale problem to that of proving the uniqueness to two simpler linear martingale problems as follows:

(i) we first establish uniqueness to the degenerate M_{k+1}-valued linear martingale problem - this uniquely determines the one dimensional marginal distributions $p_\mu(t) := \Pi_t P_\mu$ for any solution $\{P_\mu\}$ to the nonlinear martingale problem,

(ii) in the second step we establish the uniqueness to the time-inhomogeneous M_k-valued linear martingale problem associated with the operator-valued function $G_k^{\Pi_s P}$ (cf. 4.2.1).

5. UNIQUENESS TO THE MULTILEVEL POPULATION MODEL MARTINGALE PROBLEM.

Consider the M_2-valued process $\overline{\overline{X}}_{N,M}$ with generator $G_{N,M}$ given by (3.13). In this section the limit process as $N \to \infty$, $M \to \infty$, is identified. In order to describe the limiting generator consider test functions of the form

$$F_h(\nu) = \int \cdots \int \prod_{j=1}^{m} \prod_{i=1}^{n_j} h(x_{11},\ldots,x_{1,n_1};\ldots;x_{m,1},\ldots,x_{m,n_m})\mu_j(dx_{ji})\nu(d\mu_j).$$

Let

$$GF_h(\nu) = (G_1 + G_2 + G_3 + G_4)F_h(\nu) = (G_1^{\#} + G_2^{\#} + G_3^{\#} + G_4^{\#})F_\nu(h)$$

where

$$G_1^{\#}F_\nu(h) = \sum_{j=1}^{m} \sum_{i=1}^{n_j} F_\nu(A_{ji}h)$$ where A_{ji} refers to the fact that A acts on the variable x_{ji},

$G_2^{\#}F_\nu(h)$ = *individual sampling noise*

$$= \gamma_1 \sum_{j=1}^{m} \sum_{i\neq i'}^{n_j} \sum^{n_j} [F_\nu(\Phi_{ii'}^{j}h) - F_\nu(h)] ,$$

$G_3^{\#}F_\nu(h)$ = *colony sampling noise*

$$= \gamma_2 \sum_{j\neq j'}^{m} \sum^{m} [F_\nu(\Phi^{jj'}h) - F_\nu(h)] ,$$

$G_4^{\#}F_\nu(h)$ = *migration between colonies*

$$= \alpha \sum_{j=1}^{m} \sum_{i=1}^{n_j} [F_\nu(\Psi_{ji}h) - F_\nu(h)] .$$

In this model the level two fluctuations are due to colony sampling noise. If $\gamma_2 = 0$, then this is a degenerate M_2-valued model. On the other hand if $\gamma_2 \neq 0$, then this is a M_2-valued diffusion process. The case $\gamma_2 = 0$ is considered in the next theorem.

In the above expressions $\Phi^{j}_{ii'}$, $\Phi^{jj'}$ and Ψ_{ji} transform functions into functions as follows:

$$\Phi^{j}_{ii'}\, h(x_{1,1},..;x_{j,1},...,x_{j,n_{j-1}};...,x_{m,n_m})$$

$$:= h(x_{1,1},...;x_{j,1},..,x_{j,i},..,x_{j,i},..x_{j,n_j};...,x_{m,n_m})$$

$\uparrow$ *j,i'-position*

$$\Psi_{ji} h(x_{1,1},...;x_{j,1},...,x_{j,n_{j-1}};.....;,,;x_{m,1},..,x_{m,n_m};x_{m+1,1})$$

$$:= h(x_{1,1},..;x_{j,1},..,x_{m+1,1},..,x_{j,n_j};....x_{m,n_m})$$

$\uparrow$ *j,i-position*

$$\Phi^{jj'} h(x_{1,1},..;x_{j,1},..,x_{j,n_j},x_{j,n_j+1},..,x_{j,n_j+n_{j'}};...,x_{m-1,n_{m-1}})$$

j'th block $\downarrow$

$$:= h(x_{1,1},..;x_{j,1},..,x_{j,n_j};...;x_{j,n_j+1},..,x_{j,n_j+n_{j'}};,..,x_{m,n_m})$$

(jth and j'th blocks amalgamated into one block)

THEOREM 5.1. Assume that $\gamma_2 = 0$. Then the M_1-valued nonlinear martingale problem is well-posed and has a solution $X(.)$ which is a nonlinear Markov process. In addition as $N \to \infty$, and $M \to \infty$, the M_2-valued processes converge weakly to the deterministic M_2-valued process defined by $Y(t) := \Pi_t P^X$ where P^X is the probability law of the M_1-valued process $X(.)$.

Proof. We will use a bootstrap argument to prove that the M_1-valued martingale problem associated with the degenerate M_2-valued martingale problem is well-posed. The generator of the M_2-valued limiting process is

(5.1) $\quad GF_h(\nu) = (G_1 + G_2 + G_4)F_h(\nu).$

The generator of the nonlinear M_1-valued martingale problem is

(5.2) $\quad G^{\nu}F(\mu) = G_a F(\mu) + G^{\nu}_b F(\mu),$

where for test functions $F_f(\mu) = \int...\int f(x_1,...,x_n)\mu(dx_1)...\mu(dx_n),$

$$G_a F_f(\mu) = \sum_{j=1}^{n} F_\mu(A_j f) + \gamma_1 \sum_{j}^{n} \sum_{\neq k}^{n} [F_\mu(\Phi^1_{jk} f - F_\mu(f)]$$

where the right hand side is written in "dual form", and

$$\Phi^1_{jk} f(x_1,..,x_j,..,x_{n-1}) = f(x_1,..,x_j,..,x_j,..,x_n),$$

$\uparrow$ *kth position*

$$G^{\nu}_b F(\mu) = \alpha[\iint \delta F(\mu)/\delta\mu(x)\ \eta(dx)\nu(d\eta) - F(\mu)] .$$

Step 1. We first show that the G_a linear M_1-valued martingale problem is well-posed. Note that the dual form of G_a is actually the generator of a function valued process with dynamics:

(i) jumps $f \to \Phi^1_{jk} f$ at rate γ_1,

(ii) between jumps it evolves deterministically according to the semigroup associated with A.

The uniqueness to the G_a martingale problem then follows from duality (Theorem 4.1).

Step 2. We now consider the degenerate M_2-valued martingale problem associated with G given by (5.1) (with $\gamma_2 = 0$). This can be rewritten in dual form as

$$GF_h(\nu) = \sum_{j=1}^{m} F_\nu(G_a^j h) + \alpha \sum_{j=1}^{m} \sum_{i=1}^{n_j} [F_\nu(\Psi_{ji} h) - F_\nu(h)] ,$$

where G_a^j denotes the fact that the operator G_a is applied to the jth group of variables. But from Step 1 it follows that G_a generates a semigroup acting on functions of the form F_f which we denote by $\{T_t\}$. Therefore we can construct a process dual to the M_2-valued process associated with G as follows:

(i) jumps $h \to \Psi_{ji} h$ at rate $\alpha \sum_{j=1}^{m} n_j$,

(ii) between jumps it evolves according to the semigroup $T_t^1 \otimes .. \otimes T_t^m$ (tensor product).

Let $\xi(.)$ denote this dual process and $Z(t)$ denote a solution to the degenerate martingale problem in M_2 associated with G. Then by linear duality (Theorem 4.1),

$$E_{\nu_0}(F_h(Z(t))) = E_h(F_{\nu_0}(\xi(t))).$$

This means that the former problem is well-posed and also implies that for any solution P_{ν_0} of the M_1-valued nonlinear martingale problem

$$\int \ldots \int h(x_{1,1}, \ldots, x_{m,n_m}) \prod_{j=1}^{m} \prod_{i=1}^{n_j} \mu_j(dx_{ji}) \Pi_t P_{\nu_0}(d\mu_j)$$

$$= E_h(F_{\nu_0}(\xi(t)))$$

and therefore $p_{\nu_0}(t) := \Pi_t P_{\nu_0} \in M_2$ is uniquely determined for each $t \geq 0$.

Step 3. In the last step we consider time dependent linear M_1-valued martingale problem associated with

$$G^{p_{\nu_0}(t)} F(\mu) = G_a F(\mu) + \alpha[\iint \delta F(\mu)/\delta\mu(x)\, \eta(dx) p_{\nu_0}(t, d\eta) - F(\mu)].$$

We can then construct a (time-dependent) dual process $\zeta(.)$ with dynamics:

(i) jumps $f \to \Phi^1_{jk} f$ at rate γ_1,

(ii) between jumps it evolves deterministically according to the semigroup associated with A,

(iii) jumps $f \to \int f(x_1,..,x_j,..,x_n)\eta(dx_j)p_{\nu_0}(t,d\eta)$ at rate α.

Then by Theorem 4.1 this imples that the time dependent martingale problem has a unique solution. This then completes the proof that the nonlinear martingale problem is well-posed by a bootstrap argument. The proof of the weak convergence of the processes $\overline{\overline{X}}_{N,M}$ is achieved by proving uniform tightness and then showing that any limit point must satisfy the limit martingale problem which is well-posed and therefore we have weak convergence. The details are omitted.

To complete this section we state another result which can be proved using the dual processes constructed in the proof of the last theorem.

THEOREM 5.2 (SCALING LIMIT OF THE LIMIT PROCESS)

Consider the process X_α constructed by Theorem 5.1 with specified coefficient α. Assume that $A = \Delta$ (d-dim Laplacian) and let

$$Y_\varepsilon(t,B) := X_{\varepsilon^2\alpha}(t/\varepsilon^2, B_\varepsilon) \quad \text{where} \quad B_\varepsilon := \{x : \varepsilon x \in B\} .$$

As $\varepsilon \to 0$, the processes Y_ε converge weakly to a M_1-valued process with generator which can be written in dual form:

$$G_0 F_f(\mu) = G_0^{\#} F_\mu(f)$$

where

$$G_0^{\#} F_\mu(f) = F_\mu(\Delta f^*) + \alpha[F_\mu(\int\int f(x)\eta(dx)\nu(d\eta)) - F_\mu(f)]$$

and $f^*(x) = f(x,...,x)$.

COROLLARY: Let Y_0 denote the limiting process with generator G_0. Then $Y_0(t) = \delta_{b(t)}$ where $b(.)$ is a d-dimensional Brownian motion that jumps at exponentially distributed times to a new point distributed according to the probability law of $b(t)$, that is with Gaussian distribution in R^d with mean $b(0)$ and variance $2t$.

Idea of the proof. We can write the generators of the processes Y_ε as

$$G_\varepsilon^\nu F(\mu) = G_a^\varepsilon F(\mu) + G_b F(\mu)$$

where

$$G_a^\varepsilon F_f(\mu) = \sum_{j=1}^{n} F_\mu(\Delta_j f) + (\gamma_1/\varepsilon^2) \sum_{j} \sum_{\neq k}^{n} [F_\mu(\Phi_{jk}^1 f) - F_\mu(f)] .$$

Now consider the dual processes $\zeta_\varepsilon(.)$ corresponding to G_a^ε. Then there is an "averaging limit" of the processes ζ_ε and the limit is the function-valued process (living on the invariant subspace of functions of one variable) which is dual to the wandering Brownian atom $\delta_{b(.)}$. This argument shows that the transition functions for the processes Y_ε converge to the transition function for the process Y_0. The argument is completed by a tightness argument.

REFERENCES.

D.A. Dawson and K.J. Hochberg (1982). Wandering random measures in the Fleming-Viot model, Ann. Prob. 10, 554-580.

D.A. Dawson and T.G. Kurtz (1982). Application of duality to measure-valued processes, Lecture Notes in Control and Information Science 42, eds. W. Fleming and L.G. Gorostiza, Springer-Verlag, 91-105.

D.A. Dawson (1983). Critical dynamics and fluctuations for a mean-field model of cooperative behavior, J. Stat. Phys. 31, 29-85.

D.A. Dawson (1985). Asymptotic analysis of multilevel stochastic systems, Lecture Notes in Control and Information Sciences 69, eds. M. Metivier and E. Pardoux, Springer-Verlag, 79-90.

D.A. Dawson (1985). Measure-valued processes: construction, qualitative behavior and stochastic geometry, LRSP Technical Report 53, Carleton Univ. - Univ. of Ottawa.

S.M. Ermakov and A.A. Zhiglyavskii (1983). On random search for a global extremum, Th. Prob. Appl. 28, 136-141.

C. Léonard (1985). Une loi des grands nombres pour des systemes de diffusion avec interaction à coefficients non bornés, LRSP Technical Report 48, Carleton Univ. - Univ. of Ottawa.

S. Sawyer and J. Felsenstein (1982). Isolation by distance in a hierarchically clustered population, preprint.

Ya. G. Sinai (1982). Theory of Phase Transitions: Rigorous Results, Pergamon Press.

D.W. Stroock and S.R.S. Varadhan (1979). Multidimensional Diffusion Processes, Springer-Verlag Grund. 233.

V.M. Zolotarev (1984). Probability metrics, Th. Prob. Appl. 28, 278-301.

A STOCHASTIC APPROACH TO THE MINIMUM PRINCIPLE
FOR THE COMPLEX MONGE-AMPÈRE OPERATOR

M . Fukushima

§1 Introduction

We extend the minimum principle for the complex Monge-Ampère operator by employing stochastic boundary limits along the sample paths of conformal diffusions. As an application, we show that the upper regularization of the upper envelope of the Perron-Bremermann type family of subsolutions automatically satisfies the given Monge-Ampère equation inside the region.

In what follows, we denote by D an open set of C^n and by $\mathcal{P}(D)$ the family of all plurisubharmonic (psh in abbreviation) functions on D. $D_1 \Subset D$ will mean that $\bar{D}_1$ is compact and contained in D. For $z = (z^1,\cdots,z^n) \in C^n$, $z^k = x^{2k-1} + ix^{2k}$, we use the notations of differentiation

$$\frac{\partial}{\partial z^k} = \frac{1}{2}\left(\frac{\partial}{\partial x^{2k-1}} - i\frac{\partial}{\partial x^{2k}}\right), \quad \frac{\partial}{\partial \bar{z}^k} = \frac{1}{2}\left(\frac{\partial}{\partial x^{2k-1}} + i\frac{\partial}{\partial x^{2k}}\right), \quad \partial u = \sum_{k=1}^{n}\frac{\partial u}{\partial z^k}dz^k$$

$\bar{\partial}u = \sum_{k=1}^{n}\frac{\partial u}{\partial \bar{z}^k}d\bar{z}^k$ and $d = \partial + \bar{\partial}$, $d^c = i(\bar{\partial} - \partial)$. Then $dd^c u = 2i\partial\bar{\partial}u$

$= 2\sum_{k,j=1}^{n}\frac{\partial^2 u}{\partial z^k \partial \bar{z}^j} dz^k \wedge (id\bar{z}^j)$ is a differential form of bidegree (1,1) for smooth u. By definition $u \in \mathcal{P}(D)$ iff u takes values in $[-\infty,+\infty)$, $u \in L^1_{loc}(D)$, $\sum_{k,j=1}^{n}\frac{\partial^2 u}{\partial z^k \partial \bar{z}^j}\xi_k\bar{\xi}_j$ is a positive distribution (and hence a positive Radon measure) for any $\xi_k \in C$ and $u(z) = \inf_{U(z)} \operatorname{ess\,sup}_{z' \in U(z)} u(z')$ for each $z \in D$. Any psh function is R^{2n}-subharmonic but the converse is not true in general except for the case $n = 1$. For any $u \in \mathcal{P}(D)$ and open $D_1 \Subset D$, u is a decreasing limit on D_1 of C^∞-functions in $\mathcal{P}(D_1)$ ([11]).

In §2 of [2], Bedford-Taylor constructed an operator $dd^c v_1 \wedge \cdots \wedge dd^c v_k$ from $(\mathcal{P}(D) \cap L^\infty_{loc}(D))^k$ into the space of closed positive currents of bidegree (k,k), $1 \leq k \leq n$, and proved the continuity of this operation under the decreasing limits of $v_1,\cdots,v_k \in \mathcal{P}(D) \cap L^\infty_{loc}(D)$. It is this operation that we shall work with. $(dd^c v)^{\wedge k}$ is denoted by $(dd^c v)^k$. In particular, $(dd^c v)^n$ sends $v \in \mathcal{P}(D) \cap L^\infty_{loc}$ into a positive Radon measure on D and $(dd^c v)^n = 4^n \cdot n! \det\left(\frac{\partial^2 v}{\partial z^k \partial \bar{z}^j}\right) dV$ for a smooth v, V being the Lebesgue measure on D. The minimum principle for the Monge-Ampère operator $(dd^c v)^n$ reads as follows :

Minimum principle. Suppose that D is bounded and that $u, v \in \mathcal{P}(D) \cap L^\infty(D)$ satisfy

(1.1) $(dd^c u)^n \leqq (dd^c v)^n$ on D

(1.2) $\lim_{\zeta \to z} u(\zeta) \geqq \lim_{\zeta \to z} v(\zeta)$ for any $z \in \partial D$.

Then

(1.3) $u(z) \geqq v(z)$ for any $z \in D$.

This is due to Bedford-Taylor ([1 ; Theorem A] and [2 ; Corollary 4.4], cf. Cegrell [4 ; Lemma 1]).

In this paper, we are going to utilize such intimate connections between psh functions and conformal diffusions as were investigated in Schwartz[13], Fukushima [8], Okada[12] and Fukushima-Okada[9]. It was proven in [8] that any psh function u is subharmonic with respect to any conformal diffusion $M = (Z_t, \zeta, P_z)$ and moreover $u(Z_t)$ is continuous in t P_z-almost surely. Further it was shown in [8] that any pluripolar set is not hit by any conformal diffusion. In the next section, we show that those properties extend to the killing epoch ζ in the following sense.

Theorem 1. Let D be bounded and M be a conformal diffusion on D.

(i) Assume that u is psh on a neighbourhood of $\overline{D}$. Then $u(Z_t)$ is left continuous at the life time ζ :

(1.4) $\lim_{t \uparrow \zeta} u(Z_t) = u(Z_{\zeta-})$ P_z-a.s. on $\zeta < \infty$ for V-a.e. $z \in D$.

(ii) Assume that $N \subset \partial D$ and N is pluripolar. Then N is not hit by $Z_{\zeta-}$:

(1.5) $Z_{\zeta-} \in \partial D - N$ P_z-a.s. on $\zeta < \infty$ for V-a.e. $z \in \partial D$.

On the other hand, it has been proven in [9] that any closed positive current θ of bidegree $(n-1,n-1)$ with

(1.6) $\mathrm{Supp}[\theta \wedge dd^c|z|^2] = D$

gives rise to a conformal diffusion $M^\theta = (Z_t, \zeta, P_z^\theta)$ on D through a Dirichlet form $\mathcal{E}^\theta$ which will be described in §3. Keeping this in mind, take a look at the simple identity

(1.7) $(dd^c u)^n - (dd^c v)^n = (dd^c u - dd^c v) \wedge \theta$

where

(1.8) $\theta = (dd^c u)^{n-1} + (dd^c u)^{n-2} \wedge dd^c v + \cdots + (dd^c v)^{n-1}$.

θ of (1.8) is a specific closed positive current of bidegree (n-1.n-1) and the identity (1.7) would mean that $u - v$ is superharmonic with respect to the associated conformal diffusion M^{θ} provided that u, v satisfy (1.1) and θ of (1.8) satisfies (1.6).

Indeed the above reasoning will lead us in §3 to the next Theorem from which the above mentioned minimum principle follows immediately. We introduce the classes

(1.9) $\quad \mathcal{P}_{+}(D) = \{v : v(z) = w(z) + \delta|z|^2,\ v \in \mathcal{P}(D) \cap L^{\infty}(D),\ \delta > 0\}$.

(1.10) $\quad \Theta = \{\theta : \theta$ is given by (1.8) for $u \in \mathcal{P}(D) \cap L^{\infty}(D)$ and $v \in \mathcal{P}_{+}(D)\}$.

Theorem 2. Suppose that D is bounded and that $u, v \in \mathcal{P}(D) \cap L^{\infty}(D)$ satisfy (1.1) and

(1.11) $\quad \lim_{t \uparrow \zeta} u(Z_t) \geq \lim_{t \uparrow \zeta} v(Z_t) \quad P_z^{\theta}$-a.s. for V-a.e. $z \in D$ for every $\theta \in \Theta$.

Then (1.3) holds.

When $n = 1$, the closed positive current θ of bidegree (n-1, n-1) is just a constant function and E^{θ} equals the usual Dirichlet integral up to a multiplicative constant. Therefore M^{θ} is the complex Brownian motion and Theorem 2 is nothing but the classical minimum principle for the superharmonic function $u - v$ formulated in terms of Brownian motion. The moral is that in higher complex dimensions, we need a collection of diffusions $M^{\theta}, \theta \in \Theta$, rather than a single Brownian motion to formulate the minimum principle stochastically.

In §4 we combine Theorem 2 with Theorem 1 to get the next three variants of the minimum principle by modifying the condition (1.2) about the boundary values. Theorem 3 is only a slight generalization of the principle. But Theorem 4 and Theorem 5 are more useful for our present purpose and they may not be reduced to Theorem 3. In the following statements, "pluri-quasi-everywhere" or "pluri q.e." will mean "except on a pluripolar set".

Theorem 3. Assume that D is bounded and that $u, v \in \mathcal{P}(D) \cap L^{\infty}(D)$ satisfy (1.1) on D and

(1.2)' $\quad \liminf_{\substack{\zeta \to z \\ \zeta \in D}} (u(\zeta) - v(\zeta)) \geq 0$ for pluri q.e. $z \in \partial D$.

Then (1.3) holds on D.

Theorem 4. Let D be bounded open and $\tilde{D}$ be open with $D \Subset \tilde{D}$. We assume that $u \in P(D) \cap L^{\infty}(D)$, $v \in P(\tilde{D}) \cap L^{\infty}(\tilde{D})$ and they satisfy (1.1) on D and

(1.2)'' $\quad \lim_{\zeta \to z, \zeta \in D} u(\zeta) \geqq v(z)$ for pluri q.e. $z \in \partial D$.

Then (1.3) holds on D.

Theorem 5. Let D and $\tilde{D}$ be as in Theorem 4 and assume that $u, v \in P(\tilde{D}) \cap L^{\infty}(\tilde{D})$ satisfy (1.1) on D and

(1.2)''' $\quad u(z) \geqq v(z)$ for pluri q.e. $z \in \partial D$.

Then (1.3) holds on D.

Given an open set D and a function $f \in L^{\infty}_{loc}(D)$, $f \geqq 0$, we are now concerned with resolving the Monge-Ampère equation

(1.12) $$\begin{cases} u \in P(D) \cap L^{\infty}_{loc}(D) \\ (dd^c u)^n = f\, dV \quad \text{on } D \end{cases}$$

under a certain "boundary condition" by employing the Perron-Bremermann method. This method amounts to considering the class of all subsolutions

(1.13) $\quad L(f) = \{v \in P(D) \cap L^{\infty}_{loc}(D) : (dd^c v)^n \geqq f\, dV \text{ on } D\}$

and a non-empty subclass $L_o(f)$ of $L(f)$ satisfying

(1.14) $\quad v, w \in L_o(f) \Rightarrow v \vee w \in L_o(f)$

(1.15) $\quad v \in L(f)$, $w \in L_o(f)$, $v = w$ on $D - K$ for some compact $K \subset D \Rightarrow v \in L_o(f)$.

(1.16) $\quad$ the upper envelope $u_0(z) = \sup\{v(z) : v \in L_o(f)\}$, $z \in D$, is locally bounded.

Note that $L(f)$ itself satisfies condition (1.14) ([1; Prop.2.8]). Under the condition (1.16), the upper regularization u of u_o defined by $u(z) = \overline{\lim_{z' \to z}}\, u_o(z')$, becomes psh on D.

Theorem 6. Let D, f and $L_o(f)$ be as above. Then the upper regularization u of the upper envelope u_o of the class $L_o(f)$ satisfies the Monge-Ampère equation (1.12).

It has been known that, if D is bounded strongly pseudo-convex domain, $f \in L^{\infty}(D)$, $f \geqq 0$ and $\phi \in C(\partial D)$, then the Monge-Ampère equation (1.12) with the Dirichlet boundary condition

(1.17) $\quad \lim_{\zeta \to z, \zeta \in D} u(\zeta) = \phi(z)$ for any $z \in \partial D$

has a unique solution (Cegrell [4 ; Lemma 2]. This was first proven in Bedford-Taylor [1 ; Theorem D] when $f \in C(\bar{D})$. $u \in C(\bar{D})$ in that case).

In particular (1.12) and (1.17) admit a solution when D is a ball. Thus we can utilize Theorem 4, Theorem 5 and the method of the spherical modification to get Theorem 6 in §4.

We note that, if we use, instead of Theorem 4 and Theorem 5, Theorem 3 or the original form of the minimum principle only, the same proof of Theorem 6 persists to work but under an additional assumption that the upper envelope u_o be continuous (cf. [1;Theorem 8.3]). However, in the special case that $f = 0$ (the free envelope problem), the present proof of Theorem 6 does not require Theorem 4 nor Theorem 5 (cf. [2; Proposition 9.1]).

Finally we mention three corollaries of Theorem 6.

<u>Corollary 1</u>. Let D be a bounded strongly pseudo-convex domain, $f \in L^{\infty}(D)$, $f \geqq 0$, $\phi \in C(\partial D)$ and let

(1.18) $\mathcal{B}(f,\phi) = \{v \in \mathcal{P}(D) \cap L^{\infty}_{loc}(D) : (dd^c v)^n \geqq f\, dV$ on D, $\overline{\lim_{\zeta \to z}}\, v(\zeta) \leqq \phi(z)$ for any $z \in \partial D\}$.

Then $\mathcal{B}(f,\phi)$ is non-empty and the upper regularization u of the upper envelope of $\mathcal{B}(f,\phi)$ is a solution of (1.12) and (1.17).

In fact, we easily see that $\mathcal{B}(f,\phi)$ is non-empty and satisfies conditions (1.14) $\sim$ (1.16) (functions in $\mathcal{B}(f,\phi)$ are uniformly bounded because of the maximum principle for subharmonic functions). Hence u satisfies equation (1.12) by Theorem 6. Since each $z_o \in \partial D$ admits a "barrier" p_o (p_o is smooth psh in a neighbourhood of D, $p_o(z_o) = 0$, $p_o < 0$ on $\bar{D} - \{z_o\}$ and besides $(dd^c p_o)^n \geqq \delta dV$ on D for some $\delta > 0$), we see in the same way as in [3;Theorem 4.1] that u satisfies the boundary condition (1.17). The solution of (1.12) and (1.17) is unique by the minimum principle and so the function u in thr above corollary coincides with the solution constructed in [4].

<u>Corollary 2</u>. Consider a bounded (weakly) pseudo-convex domain D of the form $D = \{p < 0\}$ for some continuous psh function p defined on a neighbourhood of $\bar{D}$. Let S be the Silov boundary of D. Given a data $f \in L^{\infty}(D)$ such that

(1.19) $0 \leqq f\, dV \leqq C\, (dd^c p)^n$ on D for some $C > 0$

and a data $\phi \in C(S)$, we define $\mathcal{B}(f,\phi)$ by (1.18) but by replacing ∂D with S. Then $\mathcal{B}(f,\phi)$ is non-empty and the upper regularization u of the upper envelope of $\mathcal{B}(f,\phi)$ satisfies (1.12) and the boundary condition

(1.20) $\lim_{\zeta \to z, \zeta \in D} u(\zeta) = \phi(z)$ for any $z \in S$.

This time each point $z_o \in S$ still admits a barrier p_o but $(dd^c p_o)^n$ may degenerate ([3 ; Theorem 7.2]). However, under the condition (1.19) for f, a similar proof to the previous corollary works. (1.19) is automatically satisfied when $f = 0$. The uniqueness of the solution (1.12) under (1.20) seems to be unknown but we mention in this connection that Debiard-Gaveau [5] constructed a conformal diffusion converging to the Silov boundary S.

The next corollary concerns a characterization of the "balayage" operation, which corresponds to the case $f = 0$ and has been proven in [2].

Corollary 3. Let D be open and D_1 be an open subset with $D_1 \Subset D$. Given $\psi \in \mathcal{P}(D) \cap L^\infty_{loc}(D)$, we let

(1.21) $\mathcal{B}(D,D_1,\psi) = \{v \in \mathcal{P}(D) \cap L^\infty_{loc}(D) : v(z) \leqq \psi(z) \text{ for any } z \in D - D_1\}$.

Then the upper regularization u of the upper envelope u_o of $\mathcal{B}(D,D_1,\psi)$ is the unique function on D satisfying

$$(1.22)\quad \begin{cases} u \in \mathcal{P}(D) \cap L^\infty_{loc}(D) \\ (dd^c u)^n = 0 \quad \text{on } D_1 \end{cases}$$

$$(1.23)\quad u(z) = \psi(z) \quad \text{for pluri q.e. } z \in D - D_1.$$

In fact, the restrictions of functions in $\mathcal{B}(D,D_1,\psi)$ to D_1 satisfy condition (1.14) $\sim$ (1.16) with $f = 0$ on D_1. Hence u satisfies (1.22) by Theorem 6. Further $u = u_o$ pluri q.e. on D by virtue of [2 ; Theorem 7.1]. Since $u_o = \psi$ on $D - D_1$, we get (1.23). Let u, v be two functions satisfying (1.22) and (1.23). Then $u = v$ on D_1 by Theorem 5. Hence $u = v$ pluri q.e. on D and everywhere after all. Corollary 3 was proven in [2 ; §9]. The uniqueness part follows also from the domination principle [2 ; Corllary 4.5].

In what follows except the proof of Theorem 2(§3), we shall utilize the continuity of the operation $(dd^c v)^n$ under increasing limits of v ([2 ;Proposition 5.2]) and the $C_\#$-quasi-continuity of psh functions ([8; Proposition 3]). Actually those properties are closedly linked to the already mentioned theorem of Bedford-Taylor [2 ; Theorem 7.1] that any pluri-negligible set is pluripolar. In view of the usefulness and importance of those three properties, a new systematic proof of them in the framework of the Dirichlet space theory will be presented in a forthcoming joint paper [10].

§2 Behaviours of conformal diffusions at the killing time--Proof of Theorem 1

Let us consider a conformal diffusion $M = (Z_t, \zeta, P_z)$ on an open set $D \subset C^n$ in the sense of [8]. Equivalently, M is assumed to possess the next four properties.

(M.1) M is a strong Markov process on D and $P_z(\ Z_t$ is continuous in $t \in [0,\zeta)) = 1,\ z \in D$.

(M.2) M admits no killing inside D : $P_z(\ \tau_U < \zeta < \infty) = P_z(\zeta < \infty),\ z \in U$, for any bounded open U with $U \Subset D$. Here τ_U denotes the first leaving time from U.

(M.3) Let $\{U_\ell\}$ be open sets with $U_\ell \Subset U_{\ell+1},\ U_\ell \uparrow D$, and τ_ℓ be the first leaving time from U_ℓ. Then $P_z(\lim_{\ell\to\infty} \tau_\ell = \zeta) = 1$ for any $z \in D$.

(M.4) Let $\{\tau_\ell\}$ be as in (M.3). Then, for each ℓ, the stopped coodinate processes $Z^k_{t\wedge\tau_\ell}$, $1 \leqq k \leqq n$, and their products $Z^k_{t\wedge\tau_\ell}\cdot Z^j_{t\wedge\tau_\ell}$, $1 \leqq k, j \leqq n$, are C-valued (bounded) martingale with respect to P_z for every $z \in D$.

We then see that M enjoys the following properties :

(M.5) If D is bounded, then $Z_{\zeta-} = \lim_{t\uparrow\zeta} Z_t$ exists and $Z_{\zeta-} \in \partial D$ P_z-a.s. on $\{\zeta < \infty\}$ for $z \in D$.

(M.6) If $u \in P(D) \cap L^\infty(D)$, then $\lim_{t\uparrow\zeta} u(Z_t)$ exists P_z-a.s. for $z \in D$.

The last property is a consequence of the fact that $u(Z_{t\wedge\tau_\ell})$ is, for each ℓ, a P_z-submartingale ([8]).

Proof of Theorem 1. Let D and M be as in the statement of Theorem 1. By Proposition 2 of [8], we have for any Borel set $E \subset D$.

$$\int_D P_z(\ \sigma_E < \zeta)dV(z) \leqq C^D_\#(E)$$

where σ_E denotes the hitting time of E and

$$C^D_\#(E) = -\int_D u^*_E(z)dV(z),$$

u^*_E being the upper regularization of the upper envelope of the family

$$\{v \in P(D) : v \leqq 0 \text{ on } D,\ v \leqq -1 \text{ on } E\}.$$

Let u be psh on a neighborhood, say $\tilde{D}$, of $\bar{D}$. We may assume that $\tilde{D}$ is bounded. Because the set function $C^D_\#(\cdot)$ is monotone increasing in D, it then holds that, for any Borel set $E \subset \tilde{D}$,

$$(2.1)\qquad \int_D P_z(\ \sigma_{E\cap D} < \zeta)dV(z) \leqq C^{\tilde{D}}_\#(E)\ .$$

On the other hand, u is $C_{\#}^{\tilde{D}}$-quasi-continuous on $\tilde{D}$ by virtue of Proposition 3 of [8] : there exists a decreasing sequence of open sets $O_\ell \subset \tilde{D}$ such that $C_{\#}^{\tilde{D}}(O_\ell) \downarrow 0$ and the restriction of u to $\tilde{D} - O_\ell$ is continuous for each ℓ. In [8], this is derived by comparing $C_{\#}$ with the Bedford-Taylor capacity in [2] (denoted by $C_{B\cdot T}$) and by assuming the strong pseudo-convexity of the domain $\tilde{D}$. But, since $C_{\#}^{\tilde{D}}$ (resp. $C_{B\cdot T}^{\tilde{D}}$) is monotone increasing (resp. decreasing) in $\tilde{D}$, we need not such assumption on the domain.

By (2.1), $P_z(\bigcap_\ell \{\sigma_{O_\ell \cap D} < \zeta\}) = 0$ for V-a.e. z and consequently

(2.2) $\quad P_z(\{Z_t : t \in [0,\zeta)\} \subset \tilde{D} - O_\ell \text{ for some } \ell) = 1$ for V-a.e. $z \in D$.

Obviously this and (M.5) imply (1.4), proving the first half of Theorem 1.

To prove Theorem 1 (ii), consider a pluripolar set $N \subset \partial D$ and take a bounded open set $\tilde{D}$ containing $\bar{D}$. By definition $N \subset p^{-1}(-\infty)$ for some $p \in \mathcal{P}(C^n)$. This property is known to be equivalent to $C_{\#}^{D}(N) = 0$. On the other hand, we can see in the same way as the last part of the proof of [2 ; Proposition 6.5] that $C_{\#}^{D}$ is an outer capacity. In particular, we can find decreasing open sets O_ℓ such that $N \subset O_\ell \subset \tilde{D}$ and $C_{\#}^{D}(O_\ell) \downarrow 0, \ell \to \infty$. We then arrive at (1.5) by (2.2).

§3 A stochastic extension of the minimum principle--proof of Theorem 2

Let D be open and θ be a closed positive current of bidegree $(n-1,n-1)$ satisfying (1.6). If we set

(3.1) $\quad m = \theta \wedge dd^c|z|^2$,

then (θ, m) is an admissible pair in the sense of [9], namely, the bilinear form

(3.2) $\quad \mathcal{E}^\theta(\phi,\psi) = \int_D d\phi \wedge d^c\psi \wedge \theta, \quad \phi,\psi \in C_0^\infty(D)$,

is closable on $L^2(D;m)$ and the closure provides us with a regular Dirichlet space $(\mathcal{F}^\theta, \mathcal{E}^\theta)$ on $L^2(D;m)$ possessing the local property. Let $M^\theta = (Z_t, \zeta, P_z^\theta)$ be an associated diffusion on D : M^θ satisfies (M.1), (M.2) and (M.3) of §2 and the transition semigroup of M^θ is a realization of the L^2-semigroup generated by $(\mathcal{F}^\theta, \mathcal{E}^\theta)$.

Since θ is closed, we have for $\phi \in C_0^\infty(D)$ and $\psi \in C^\infty(D)$ $(\subset \mathcal{F}^\theta_{loc})$

(3.3) $\quad \mathcal{E}^\theta(\phi, \psi) = -\int_D \phi \, dd^c\psi \wedge \theta$.

This also holds for complex valued C^∞-function ψ and, in particular, the coodinate functions $\psi(z) = z^k$ and their products $\psi(z) = z^k \bar{z}^j$, $1 \leq k, j \leq n$, are $\mathcal{E}^\theta$-harmonic : $\mathcal{E}^\theta(\phi, \psi) = 0$ for any $\phi \in C_0^\infty(D)$. This means that M^θ meets the condition (M.4) of §2 for $\mathcal{E}^{\theta}$-q.e. $z \in D$ ([9 ; pp382]). By modifying every point in a (proper) exceptional set to be trap, we can achieve (M.4) holding for every $z \in D$. Thus we may assume that M^{θ} is a conformal diffusion in the sense of §2.

Lemma 1

(i) If D is bounded, then

(3.4) $P_z^{\theta}(\zeta < \infty) = 1$ for $\mathcal{E}^{\theta}$-q.e. $z \in D$.

(ii) Suppose $w \in \mathcal{F}^{\theta}_{loc}$, w is $\mathcal{E}^{\theta}$-quasi-continuous and w is $\mathcal{E}^{\theta}$-superharmonic :

(3.5) $\mathcal{E}^{\theta}(\phi\ , w) \geqq 0$ for any $\phi \in C_0^{\infty}(D)$, $\phi \geqq 0$.

Then w is M^{θ}-superharmonic : for any open $G \Subset D$,

(3.6) $w(z) \geqq E_z^{\theta}(w(Z_{\tau_G}))$ for $\mathcal{E}^{\theta}$-q.e. $z \in D$.

Proof (i) can be shown exactly in the same manner as in the proof of Lemma 6 of [9]. Turning to the proof of (ii), we first note that, by (3.5), there exists a unique positive Radon measure μ on D such that

(3.7) $\mathcal{E}^{\theta}(\phi\ , w) = \int_D \phi\, d\mu$, $\phi \in C_0^{\infty}(D)$.

We claim that μ charges no set of zero capacity (and consequently μ is a smooth measure in the sense of [6]) and that, for any open $G \Subset D$,

(3.8) $\mathcal{E}^{\theta}(\ \phi, w) = \int_D \tilde{\phi}(x)\mu(dx)$, $\phi \in \mathcal{F}_G^{\theta}$,

where $\mathcal{F}_G^{\theta} = \{\ \phi \in \mathcal{F}^{\theta}: \tilde{\phi} = 0\ \ \mathcal{E}^{\theta}\text{-q.e. on } D - G\}$,$\tilde{\phi}$ being an $\mathcal{E}^{\theta}$-quasi-continuous modification of ϕ.

We remark that the restriction $\mathcal{E}_G^{\theta}$ of $\mathcal{E}^{\theta}$ to $\mathcal{F}_G^{\theta}$ is a Dirichlet form on $L^2(G;m)$, which is C_0^{∞}-regular by the spectral synthesis ([6 ; Theorem 3.3.4 and Problem 3.3.4]). Further a function on G (resp. a subset of G) is quasi-continuous (resp. of zero capacity) with respect to $\mathcal{E}^{\theta}$ if and only if it is so with respect $\mathcal{E}_G^{\theta}$ ([6 ; Theorem 4.4.2]). Take now $w_1 \in \mathcal{F}^{\theta}$ such that $w = w_1$ on G. We have then from (3.7), for any $\phi \in C_0^{\infty}(G)$,

$$\int |\phi|\, d\mu = \mathcal{E}^{\theta}(|\phi|, w_1) \leqq \sqrt{\mathcal{E}^{\theta}(w_1, w_1)}\sqrt{\mathcal{E}^{\theta}(\phi,\phi)},$$

which means that $\mu|_G$ is of finite energy integral with respect to $\mathcal{E}_G^{\theta}$. Hence $\mu|_G$ charges no set of $\mathcal{E}_G^{\theta}$-capacity zero and (3.8) for $\mathcal{E}_G^{\theta}$-quasi-continuous version $\tilde{\phi}$ follows from (3.7) just in the same way as in the proof of [6 ;Theorem 3.2.2].

Let A be the positive continuous additive functional of M^{θ} associated with the smooth measure μ. Since (3.8) holds for any open set $G \Subset D$, Theorem A2 of [7] applies and we get

(3.9) $w(Z_t) - w(Z_0) = M_t^{[w]} - A_t$, $t < \zeta$, P_z-a.s. for q.e. $z \in D$.

Here $M_t^{[w]}$ is a local martingale additive functional such that for any open $\tilde{G} \Subset D$ and for $w_1 \in F^\theta$ with $w = w_1$ on $\tilde{G}$, $M_t^{[w]}$ equals the square integrable mean zero additive functional $M_t^{[w_1]}$ for $t < \tau_{\tilde{G}}$.

The desired inequality (3.6) is an immediate consequence of (3.9). In fact, for open $G, \tilde{G}$ with $G \Subset \tilde{G} \Subset D$, we have $\tau_G < \tau_{\tilde{G}} < \zeta < +\infty$ P_z-a.s. for E^θ-q.e. $z \in G$ by (M.2) and (3.4). Hence $E_z^\theta(M_{\tau_G}^{[w_1]}) = 0$ by the optional sampling theorem for the martingale $M_t^{[w_1]}$ and consequently $E_z^\theta(w(Z_{\tau_G})) - w(z) = - E_z^\theta(A_{\tau_G})$ $\leqq 0$ for E^θ-q.e. $z \in G$. Since $P_z^\theta(\tau_G = 0) = 1$ for E^θ-q.e. $z \in D - G$ by [6; Theorem 4.2.3], the proof of Lemma 1 is now complete.

Now we assume that D is bounded and we take θ from the family Θ of (1.10) of §1. Obviously θ meets the condition (1.6) and indeed m of (3.1) dominates the Lebesgue measure :

(3.10) $m \geq \delta' V$ for some $\delta' > 0$.

Let F^θ, E^θ, $M^\theta = (Z_t, \zeta, P_z^\theta)$ be the associated objects.

<u>Lemma 2</u> Any $w \in P(D) \cap L_{loc}^\infty(D)$ is in F_{loc}^θ and E^θ-quasi-continuous.

Moreover

(3.11) $E^\theta(\phi, w) = -\int_D \phi \, dd^c w \wedge \theta$, $\phi \in C_o^\infty(D)$.

<u>Proof</u> The proof is essentially the same as Lemma 7 of [9]. But we give a detailed proof of the first assertion for completeness. Take any open $D_1 \Subset D$ and $\phi \in C_o^\infty(D_1)$. Take $w_k \in P(D_1) \cap C^\infty(D_1)$ which decreases to w on D_1. Regarding ϕw_k as elements of $C_o^\infty(D)$, we have, by the Schwarz type inequality [12 ; Lemma 1], $E^\theta(\phi w_k - \phi w_\ell, \phi w_k - \phi w_\ell)$

$$\leqq 2\int_{D_1} (w_k - w_\ell)^2 d\phi \wedge d^c\phi \wedge \theta + 2\int_{D_1} \phi^2 d(w_k - w_\ell) \wedge d^c(w_k - w_\ell) \wedge \theta .$$

The right hand side converges to zero as $k, \ell \to \infty$ by virtue of [2 ; Theorem 2.7]. Since $\phi w_k \to \phi w$ pointwise and in $L^2(D;m)$ as well, we conclude that ϕw is an E^θ-quasi-continuous function in F^θ and ϕw_k is E_1^θ-convergent to ϕw.

Proof of Theorem 2 Let D be bounded and $u, v \in \mathcal{P}(D) \cap L^{\infty}(D)$ satisfy (1.1) and (1.11). Let, for $\delta > 0$,

(3.12) $$v^{(\delta)} = v + \delta|z|^2 - \delta\gamma \quad \text{with } \gamma = \sup_{z \in D} |z|^2 ,$$

then the pair u and $v^{(\delta)}$ still satisfy (1.1) and (1.11). Note that the stochastic limits in (1.11) always exist in view of (M.6) of §2.

Fix $\delta > 0$ and consider θ of (1.8) defined for this particular pair u and $v^{(\delta)}$. Since θ belongs to the family Θ of (1.10), we may think of the associated objects $\mathcal{F}^{\theta}$, $\mathcal{E}^{\theta}$ and $M^{\theta} = (Z_t, \zeta, P_z^{\theta})$. By virtue of Lemma 2 and the identity (1.7), we see that $u - v^{(\delta)}$ is an $\mathcal{E}^{\theta}$-quasi-continuous function in $\mathcal{F}^{\theta}_{loc}$ and, for any $\phi \in C_o^{\infty}(D)$,

(3.13) $$\mathcal{E}^{\theta}(\phi, u - v^{(\delta)}) = -\int_D \phi \{(dd^c u)^n - (dd^c v^{(\delta)})^n\} .$$

Now (3.13) and inequality (1.1) holding for u and $v^{(\delta)}$ lead us to the conclusion that $u - v^{(\delta)}$ is $\mathcal{E}^{\theta}$-superharmonic in the sense of (3.5). By Lemma 1, $u(z) - v^{(\delta)}(z) \geqq E_z^{\theta}((u - v^{(\delta)})(Z_{\tau_G}))$ for $\mathcal{E}^{\theta}$-q.e. $z \in D$ for any open $G \Subset D$. By letting $G \uparrow D$, $u(z) - v^{(\delta)}(z) \geqq E_z^{\theta}(\lim_{t \uparrow \zeta} \{u(Z_t) - v^{(\delta)}(Z_t)\})$, which holds $\mathcal{E}^{\theta}$-q.e. $z \in D$ and hence m-a.e., and V-a.e. according to (3.10). Since u and $v^{(\delta)}$ satisfy (1.11), we get $u(z) \geqq v^{(\delta)}(z)$ for V-a.e. $z \in D$ and consequently for every $z \in D$. Finally we arrive at (1.3) by letting $\delta \downarrow 0$.

§4 Analytical consequences--proof of Theorems 3,4,5 and 6

For D bounded and for $\theta \in \Theta$, $M^{\theta} = (Z_t, \zeta, P_z^{\theta})$ satisfies

(4.1) $$P_z^{\theta}(\zeta < +\infty, Z_{\zeta-} \in \partial D) = 1 \quad \text{V-a.e. } z \in D,$$

by virtue of (M.5), Lemma 1 (i) and (3.10).

Proof of Theorem 3 Condition (1.2)' of Theorem 3 implies the conditon (1.11) of Theorem 2 in view of (4.1) and Theorem 1 (ii). Hence Theorem 3 is a consequence of Theorem 1 (ii) and Theorem 2.

Proof of Theorem 4 In view of (4.1), the condition (1.2)'' of Theorem 4 coupled with Theorem 1 implies condition (1.11) of Theorem 2. Therefore Theorem 4 is a consequence of Theorem 1 and Theorem 2.

Proof of Theorem 5 Similar to the proof of Theorem 4.

Proof of Theorem 6 Take any $v \in \mathcal{L}_o(f)$ and any open balls B, B_1 such that $B \Subset B_1 \Subset D$. There exist then bounded $v_k \in \mathcal{P}(B_1) \cap C^\infty(B_1)$, $k = 1,2,\cdots$, which decrease to v on B_1. By Cegrell [4 ; Lemma 2], there exists $w_k \in \mathcal{P}(B) \cap L^\infty(B)$ solving the Monge-Ampère equation on B with boundary value v_k :

$$(dd^c w_k)^n = f\,dV \text{ on } B, \quad \lim_{\substack{\zeta \to z \\ \zeta \in B}} w_k(\zeta) = v_k(z) \quad z \in \partial B.$$

w_k is decreasing on B by Theorem 3 . Furthermore, since $(dd^c w_k)^n \leq (dd^c v)^n$ on B and $\lim_{\substack{\zeta \to z \\ \zeta \in B}} w_k(\zeta) \geq v(z)$ for any $z \in \partial B$, we can make use of Theorem 4 to conclude that $w_k \geq v$ on B.

Let $\lim_{k\to\infty} w_k(z) = w(z)$, $z \in B$, then $w \in \mathcal{P}(B) \cap L^\infty(B)$, $w \geq v$ on B and $(dd^c w)^n = f\,dV$ on B. We perform a spherical modification of v by

$$\tilde{v} = \begin{cases} w & \text{on } B \\ v & \text{on } D - B. \end{cases}$$

$\tilde{v}$ is then upper semi-continuous on D, $\tilde{v} \geq v$ on D and $\tilde{v} = v$ on $D - B$ and hence $\tilde{v} \in \mathcal{P}(D) \cap L^\infty_{loc}(D)$ according to the fact that a finite valued function on D is psh if and only if it is upper semi-continuous and its restriction on each complex straight line is R^2-subharmonic ([11]). We further have $\tilde{v} \in \mathcal{L}(f)$ because $(dd^c\tilde{v})^n$ is equal to fdV on B, greater than fdV on $D - \bar{B}$ and the Lebesgue measure does not charge on ∂B. Hence $\tilde{v} \in \mathcal{L}_o(f)$ in view of the condition (1.15). Summing up, we have constructed, for each $v \in \mathcal{L}_o(f)$, a function $\tilde{v} \in \mathcal{L}_o(f)$ such that

(4.2) $\tilde{v} \geq v$ on D, $\tilde{v} = v$ on $D - B$, $(dd^c\tilde{v})^n = f\,dV$ on B.

Let u be the upper regularization of the upper envelope u_o of the class $\mathcal{L}_o(f)$. By Choquet's lemma, we can find $u_j \in \mathcal{L}_o(f)$ such that u_j is non-decreasing and $\lim_{j\to\infty} u_j(z) = u(z)$ V-a.e. Denote by $\tilde{u}_j$ the spherical modification of u_j as above. Then, for $j > k$, $(dd^c\tilde{u}_j)^n = (dd^c\tilde{u}_k)^n$ on B and $\tilde{u}_j(z) = u_j(z) \geq u_k(z) = \tilde{u}_k(z)$, $z \in \partial B$. Hence, this time we can use Theorem 5 to conclude that $\tilde{u}_j$ is non-decreasing again.

Since $\tilde{u}_j \in \mathcal{L}_o(f)$, we have $u_j \leq \tilde{u}_j \leq u_o$ and consequently $\lim_{j\to\infty} \tilde{u}_j(z) = u(z)$ V-a.e. We now apply [2 ; Proposition 5.2] to $\tilde{u}_j$ and u and we get $(dd^c u)^n = \lim_{j\to\infty} (dd^c\tilde{u}_j)^n = fdV$ on B. B being arbitrary, u satisfies the Monge-Ampère equation (1.12).

References

[1] E. Bedford and B.A. Tayor, The Dirichlet problem for a complex Monge-Ampère equation, Inventiones Math. 37 (1976), 1-44.

[2] E. Bedford and B.A. Taylor, A new capacity for plurisubharmonic functions, Acta Math. 149 (1982), 1-40.

[3] H.J. Bremermann, On a generalized Dirichlet problem for plurisubharmonic functions and pseudo-convex domains, characterization of Silov boundaries, Trans.Amer. Math. Soc. 91(1959), 246-276.

[4] U. Cegrell, on the Dirichlet problem for the complex Monge-Ampère operator, Math. Z. 185(1984), 247-251.

[5] A. Debiard et B. Gaveau,Frontiere de Silov de domaines faiblement pseudo-convexes de C^n, Bull. Sc. Math., 2e serie, 100(1976), 17-31.

[6] M. Fukushima, Dirichlet forms and Markov processes, North Holland,Amsterdam, 198o.

[7] M. Fukushima, On absolute continuity of multidimensional symmetrizable diffusions, in Lecture Notes in Math., vol.923, Springer, 1982.

[8] M. Fukushima, On the continuity of plurisubharmonic functions along conformal diffusions, Osaka J. Math. to appear.

[9] M. Fukushima and M. Okada, On conformal martingale diffusions and pluripolar sets, J. Func, Anal. 55(1984), 377-388.

[10] M. Fukushima and M. Okada, On Dirichlet forms for plurisubharmonic functions, to appear.

[11] P. Lelong, Plurisubharmonic functions and positive differential forms, Gordon and Breach, New York, 1969.

[12] M. Okada, Espaces de Dirichlet généraux en analyse complexe, J. Func.Anal. 46(1982), 396-410.

[13] L. Schwartz, Semi-martingales sure des varietés, et martingales conformes sur des varietés analytiques complexes, Lecture Notes in Math. Vol.780,Springer, 1980.

Masatoshi Fukushima
Department of Mathematics
College of General Education
Osaka University
Toyonaka, Osaka

CONSTRUCTION OF STOCHASTIC PROCESSES ASSOCIATED WITH THE BOLTZMANN EQUATION AND ITS APPLICATIONS

Tadahisa Funaki
Department of Mathematics
Faculty of Science
Nagoya University
Nagoya, 464, JAPAN

1. Introduction.

The nonlinear Boltzmann equation in the kinetic theory of rarefied gases has been studied probabilistically in the spatially homogeneous case by several authors including Tanaka [18], Sznitman [17] and Funaki [4]. They constructed a class of jump processes with mean-field interaction of the type introduced by McKean [11]. The purpose of this note is twofold. The first is an extension of their methods, more precisely, the construction of jump processes in the spatially inhomogeneous case. The second is to give a microscopic interpretation in terms of jump processes of the derivation of the compressible Euler equation from the Boltzmann equation.

The Boltzmann equation is written as

$$(1.1)\quad \begin{cases} \dfrac{\partial f}{\partial t} + \xi\cdot\nabla_x f = Q[f,f] \quad , & (t,x,\xi) \in (0,T]\times R^3\times R^3 \quad , \quad T>0, \\ f(0) = f_0 \quad , & (x,\xi) \in R^3\times R^3. \end{cases}$$

Here $f = f(t) = f(t,x,\xi)$ is the distribution density of gas particles with position x and velocity ξ at time t and Q is a collision integral acting only on the variable ξ given by

$$Q[f,f_1](\xi) = \int_{(0,\pi/2)\times(0,2\pi)\times R^3} \{f(\xi')f_1(\xi_1') - f(\xi)f_1(\xi_1)\}\, B(\theta,|\xi-\xi_1|)d\theta d\phi d\xi_1,$$

where

$$\xi' = \xi + \alpha(\alpha,\xi_1-\xi) \quad , \quad \xi_1' = \xi_1 - \alpha(\alpha,\xi_1-\xi)$$

$$(\alpha,\xi_1-\xi) = |\xi_1-\xi|\cos\theta \quad , \quad \alpha\in R^3 \,;\, |\alpha| = 1 \quad , \quad \sin\theta d\theta d\phi = d\alpha.$$

Throughout the paper we put the angular cutoff assumption of Grad [6], i.e., $B(\theta,|\xi|)$ is continuous in $|\xi| \in (0,\infty)$ and has a bound:

$$0 \leq B(\theta,|\xi|) \leq C(|\xi| + |\xi|^{-\delta}) \quad , \quad C > 0 \quad , \quad 0 \leq \delta < 3.$$

For a given function $f_1 = f_1(\xi)$ on R^3, a formal dual operator $\mathcal{B}_{f_1}$

of $Q[\cdot,f_1]$ acting on the space $L^2(R^3,d\xi)$ can be expressed as

$$\mathcal{B}_{f_1}\psi(\xi) = \int_{(0,\pi/2)\times(0,2\pi)\times R^3} \{\psi(\xi')-\psi(\xi)\}B(\theta,|\xi-\xi_1|)f_1(\xi_1)d\theta d\phi d\xi_1, \quad \psi\in C_0^\infty(R^3).$$

The initial data f_0 in (1.1) is a nonnegative function on $R^3\times R^3$ and we discuss the following two cases separately.

(I) f_0 is a probability density, i.e., $\int_{R^3\times R^3} f_0(x,\xi)dxd\xi = 1$,

(II) f_0 is not integrable, i.e., $\int_{R^3\times R^3} f_0(x,\xi)dxd\xi = \infty$.

In the case (I) the construction problem of stochastic processes associated with the Boltzmann equation is formulated as the following martingale problem: Find a probability measure P on the path space $D([0,T],R^3\times R^3)$, with a coordinate function $\{y_t=(x_t,\xi_t),\ t\in[0,T]\}$ and a usual increasing family of σ-fields $\{\mathcal{F}_t,\ t\in[0,T]\}$, which satisfies

[P.I]
- (i) For each $t\in[0,T]$, $P\circ y_t^{-1}$ is absolutely continuous with respect to the Lebesgue measure $dxd\xi$ on $R^3\times R^3$ and has a density $f(t,x,\xi)$ (i.e., $P(y_t\in dxd\xi) = f(t,x,\xi)dxd\xi$),
- (ii) $f(0,x,\xi) = f_0(x,\xi)$ a.e.,
- (iii) for each $\psi\in C_0^\infty(R^3\times R^3)$,

$$\psi(y_t) - \int_0^t \mathcal{L}_{f(s)}\psi(y_s)ds \quad , \quad 0 \leqq t \leqq T,$$

is a martingale relative to $(P,\{\mathcal{F}_t\})$,

where $\mathcal{L}_f$, $f = f(x,\xi)$, is an operator defined by

$$\mathcal{L}_f\psi(x,\xi) = \xi\cdot\nabla_x\psi(x,\xi) + \{\mathcal{B}_{f(x,\cdot)}\psi(x,\cdot)\}(\xi).$$

While in the case (II) our goal is to construct an infinite system of stochastic processes $\{y_t^i=(x_t^i,\xi_t^i),\ t\in[0,T]\}_{i=1}^\infty$ with $y_\cdot^i\in D([0,T],R^3\times R^3)$ on a proper probability space $(\Omega,\mathcal{F},P)$ equipped with a reference family $\{\mathcal{F}_t,\ t\in[0,T]\}$ in such a way that they satisfy the following condition:

[P.II]
- (0) For each $i = 1,2,\cdots$, the process y_t^i is $\{\mathcal{F}_t\}$-adapted and under the conditional probability distribution of P given $\mathcal{F}_0$ $\{y_\cdot^i\}_{i=1}^\infty$ is an independent system,
- (i) for each $t\in[0,T]$ and $i = 1,2,\cdots$, $P\circ(y_t^i)^{-1}$ is absolutely continuous with respect to $dxd\xi$ and

$$f(t,x,\xi) = E[\#\{i\,|\,y_t^i\in dxd\xi\}]/dxd\xi < \infty \quad \text{a.e.},$$

- (ii) $f(0,x,\xi) = f_0(x,\xi)$ a.e.,

(iii) for each $i = 1,2,\cdots$ and $\psi \in C_0^\infty(R^3\times R^3)$,

$$\psi(y_t^i) - \int_0^t L_{f(s)}\psi(y_s^i)ds \quad , \quad 0 \leq t \leq T,$$

is a martingale relative to $(P,\{F_t\})$.

In either case (I) or (II), if we can construct the stochastic process, then it is clear that f solves a weak version of the Boltzmann equation:

(1.2) $$\frac{d}{dt}\int_{R^3\times R^3}\psi(x,\xi)f(t,x,\xi)dxd\xi = \int_{R^3\times R^3}\{\xi\cdot\nabla_x\psi f(t)+Q[f(t),f(t)]\psi\}dxd\xi$$

for every $\psi \in C_0^\infty(R^3\times R^3)$.

In Section 2, we investigate a linear forward equation arrising from the Boltzmann equation and prove the conservation law of mass for solutions by using a fundamental estimate due to Grad, namely, it is shown that the (minimal) solution to this linear equation is a probability density if its initial data is. This implies the nonexplosion property of solutions to an $\{L_{f_1(t)}\}$-martingale problem with given $f_1(t)$ (see Section 3). The construction of stochastic processes will be given in Section 4. Our assumption, which is guaranteed by several analytical results, is that the Boltzmann equation (1.1) has a nonnegative solution $f(t,x,\xi)$ satisfying

$$\sup\{(1+|\xi|)g^{-1/2}(\xi)f(t,x,\xi)\,;\, t \in [0,T],\ x,\xi \in R^3\} < \infty,$$

where $g(\xi) = (2\pi)^{-3/2}\exp\{-|\xi|^2/2\}$ is a so-called Maxwellian distribution. Finally in Section 5, assuming some additional condition, the derivation of the compressible Euler equation will be discussed and formulated as a law of large numbers for an $S'(R^3)$-valued process $\mu_t^\varepsilon = \sum_{i=1}^\infty \delta_{x_t^{i,\varepsilon}}$, where $\{x_t^{i,\varepsilon}\}_{i=1}^\infty$ is a family of stochastic processes which are associated with the Boltzmann equation with mean free path $\varepsilon > 0$, i.e., the equation (1.1) with Q replaced by $\frac{1}{\varepsilon}Q$.

2. Linear forward equation.

For each measurable function $v = v(\xi)$ on R^3 satisfying $\|v\|_1 = \sup_\xi (1+|\xi|)|v(\xi)| < \infty$, an operator H_v on the space $L^\infty(R^3)$ is introduced by

$$\{H_v u\}(\xi) = \int u(\xi')v(\xi_1')\omega(\xi_1)B(\theta,|\xi-\xi_1|)d\theta d\phi d\xi_1, \quad u \in L^\infty(R^3),$$

where $\omega(\xi) = \sqrt{g(\xi)}$. We sometimes omit the domain of integration when it is clear. A slight modification of the method used by Grad [6] gives the following estimate with a positive constant C:

(2.1) $\qquad \|H_v u\|_\infty \leq C\,\|v\|_1\,\|u\|_\infty \quad , \quad u \in L^\infty(R^3).$

We shall denote by $\|\cdot\|_\infty$ the usual supremum-norm of functions on R^3 or $R^3\times R^3$ without distinction.

Let $u_0 = u_0(x,\xi)$ and $u_1 = u_1(t,x,\xi)$, $(t,x,\xi) \in R_+\times R^3\times R^3$, $R_+ = [0,\infty)$, be nonnegative bounded measurable functions. The function u_1 is assumed to be continuous in (x,ξ) and satisfy $\sup_{t,x}\|u_1(t,x,\cdot)\|_1 < \infty$. We put

$$\bar{f}_0(x,\xi) = \omega(\xi)u_0(x,\xi), \quad f_1(t,x,\xi) = \omega(\xi)u_1(t,x,\xi),$$
$$\nu(t,x,\xi) = \int f_1(t,x,\xi_1)B(\theta,|\xi-\xi_1|)d\theta d\phi d\xi_1,$$

and define bounded operators $U(t,s)$ and H_t , $0\leq s\leq t$, on the space $L^\infty(R^3\times R^3)$ as follows:

$$\{U(t,s)u\}(x,\xi) = \exp\{-\int_s^t \nu(r,x-\xi(t-r),\xi)dr\}u(x-\xi(t-s),\xi),$$
$$\{H_t u\}(x,\xi) = \{H_{u_1(t,x,\cdot)}u(x,\cdot)\}(\xi) \quad , \quad u \in L^\infty(R^3\times R^3).$$

Then the estimate (2.1) shows that a series $u(t) = \sum_{k=0}^{\infty} u_k(t)$ converges in the space $L^\infty(R^3\times R^3)$, where nonnegative functions $u_k(t)$ are defined inductively by

$$\text{(2.2)} \quad \begin{cases} u_0(t) = U(t,0)u_0 \\ u_{k+1}(t) = \int_0^t U(t,s)H_s u_k(s)ds \quad , \quad k = 0,1,2,\cdots. \end{cases}$$

See Section 3 for a probabilistic meaning of this series. The following lemma can be proven without difficulty.

Lemma 2.1 (i) The function $u(t)$, $t \geq 0$, is a unique solution to the following integral equation on the space $L^\infty(R^3\times R^3)$ satisfying $\sup_{0\leq t\leq T}\|u(t)\|_\infty < \infty$, $T > 0$:

$$\text{(2.3)} \quad u(t) = U(t,0)u_0 + \int_0^t U(t,s)H_s u(s)ds, \quad t \geq 0.$$

(ii) The function $f(t,x,\xi) = \omega(\xi)u(t,x,\xi)$ satisfies a weak linear forward equation:

$$\text{(2.4)} \quad \langle f(t),\psi\rangle = \langle \bar{f}_0,\psi\rangle + \int_0^t \langle f(s), L_{f_1(s)}\psi\rangle ds \quad , \quad \psi \in C_0^\infty(R^3\times R^3),$$

where L is the operator introduced in Section 1 and

$$\langle f,\psi\rangle = \int_{R^3\times R^3} f(x,\xi)\psi(x,\xi)dxd\xi \ .$$

Remark 2.1 The strong version of the equation (2.4) is

$$(2.4)' \quad \begin{cases} \frac{\partial f}{\partial t} + \xi\cdot\nabla_x f = Q[f,f_1] \quad, \quad (t,x,\xi) \in (0,\infty)\times R^3\times R^3, \\ f(0) = \overline{f}_0. \end{cases}$$

Next we state the conservation law of mass:

Lemma 2.2 If the function $\overline{f}_0$ is integrable, then we have

$$\int_{R^3\times R^3} f(t,x,\xi)dxd\xi = \int_{R^3\times R^3} \overline{f}_0(x,\xi)dxd\xi, \quad t \geq 0.$$

Proof. We may only complete the proof in the case where $\overline{f}_0$ has a compact support. Since the function $v(\xi) = \sup_{t,x} u_1(t,x,\xi)$ satisfies $\|v\|_1 < \infty$, we have

$$\begin{aligned} \int_{R^3} u_{k+1}(t,x,\xi)dx &\leq \int_0^t ds\int_{R^3} \{H_s u_k(s)\}(x-\xi(t-s),\xi)dx \\ &\leq \int_0^t ds\int_{R^3} \{H_v u_k(s)\}(x,\xi)dx \\ &= \int_0^t ds\ H_v \{\int_{R^3} u_k(s,x,\cdot)dx\}(\xi) \\ &\leq C\,\|v\|_1 \int_0^t \|u_k(s)\|ds, \end{aligned}$$

for every $\xi \in R^3$, where we put

$$\|u\| = \sup_\xi \int_{R^3} u(x,\xi)dx.$$

Hence we obtain

$$\|u_{k+1}(t)\| \leq C\,\|v\|_1\int_0^t \|u_k(s)\|\,ds \quad, \quad k = 0,1,2,\cdots,$$

which implies

$$(2.5) \qquad \|u_k(t)\| \leq \{C\,\|v\|_1 t\}^k\|u_0\|/k! \quad < \infty .$$

Now by differentiating the function

$$\begin{aligned} &\int\{U(t,s)H_s u_{k-1}(s)\}(x,\xi)\omega(\xi)dxd\xi \\ &= \int\exp\{-\int_s^t \nu(r,x+\xi(r-s),\xi)dr\}\{H_s u_{k-1}(s)\}(x,\xi)\omega(\xi)dxd\xi, \quad t\geq s\geq 0, \end{aligned}$$

in t, we can prove

$$\begin{aligned} \int u_k(t,x,\xi)\omega(\xi)dxd\xi = &\int_0^t ds\int\nu(s,x,\xi)u_{k-1}(s,x,\xi)\omega(\xi)dxd\xi \\ &- \int_0^t ds\int\nu(s,x,\xi)u_k(s,x,\xi)\omega(\xi)dxd\xi, \quad k = 1,2,\cdots, \end{aligned}$$

and same equality for $k = 0$ with the first term in the right hand side replaced by $\int \overline{f}_0(x,\xi)dxd\xi$. We should note in these calculations that all integrals converge absolutely because of the bound (2.5) and the following estimate (see Grad [6]):

$$\nu(s,x,\xi) \leqq \text{const.}(1+|\xi|) \quad \text{for every} \quad (s,x,\xi).$$

We therefore obtain

$$\int \sum_{k=0}^{N} u_k(t,x,\xi)\omega(\xi)dxd\xi - \int \overline{f}_0(x,\xi)dxd\xi$$
$$= \int_0^t ds \int \nu(s,x,\xi)u_N(s,x,\xi)\omega(\xi)dxd\xi,$$

which implies the conclusion since (2.5) shows that the right hand side converges to 0 as N tends to ∞. □

3. Linear martingale problem.

Take a function $f_1 = f_1(t,x,\xi)$ and a probability density $\overline{f}_0 = \overline{f}_0(x,\xi)$ as in Section 2. The purpose of this section is to prove that no explosion occurs for solutions to the martingale problem with infinitesimal operators $\{L_{f_1(t)}, t \geqq 0\}$. Note that the total jump rate $\nu(t,x,\xi)$ is generally unbounded.

Let $S = R^3 \times R^3 \cup \{\Delta\}$ be a one-point compactification of $R^3 \times R^3$ and let $P = P_{s,x,\xi}$, $(s,x,\xi) \in R_+ \times R^3 \times R^3$, be a probability measure on $D([s,\infty),S)$ satisfying the following condition:

$$(3.1) \quad \begin{cases} \text{(i)} \quad P(y_s = (x,\xi)) = 1, \\ \text{(ii) for every } \psi \in C_0^\infty(R^3 \times R^3) \text{ and } n = 1,2,\cdots, \\ \qquad \psi(y_{t\wedge\tau_n}) - \int_s^{t\wedge\tau_n} L_{f_1(r)}\psi(y_r)dr \quad , \quad t \geqq s, \\ \qquad \text{is a martingale relative to } (P,\{F_t\}), \\ \text{(iii)} \quad P(y_t = \Delta, t \geqq e) = 1, \end{cases}$$

where $\{\tau_n\}_{n=0}^\infty$ is a sequence of stopping times defined inductively by

$$\tau_n = \inf\{ t > \tau_{n-1} ; y_t \neq y_{t-} \} , \quad n \geqq 1 \ ; \ \tau_0 = s ,$$

and

$$e = \lim_{n\to\infty} \tau_n \in (s,\infty].$$

The method of time change (see Ikeda and Watanabe [8]) or a usual method to construct transport processes (see Papanicolaou [14]) proves the existence and uniqueness of solutions $P_{s,x,\xi}$ to the martingale problem (3.1) for each (s,x,ξ).

Let $f(t,x,\xi) = \sum_{k=0}^{\infty} \omega(\xi)u_k(t,x,\xi)$ be the function given in

Section 2. Then we can prove

$$(3.2)\qquad E^{P_{0,\bar{f}_0}}[\psi(y_t)\,;\,\tau_k \leqq t < \tau_{k+1}] = \int \psi(x,\xi)\omega(\xi)u_k(t,x,\xi)dxd\xi,$$

for every $t \geqq 0$, $k = 0,1,2,\cdots$ and $\psi \in C_b(R^3\times R^3)$. Here we put

$$(3.3)\qquad P_{0,\bar{f}_0}(\cdot) = \int P_{0,x,\xi}(\cdot)\bar{f}_0(x,\xi)dxd\xi.$$

An immediate consequence of (3.2) combined with Lemma 2.2 is the following:

$$(3.4)\quad \begin{cases} \text{(i)} & P_{0,\bar{f}_0}(e=\infty) = 1, \\ \text{(ii)} & P_{0,\bar{f}_0}(y_t \in dxd\xi) = f(t,x,\xi)dxd\xi \quad , \quad t \geqq 0, \\ \text{(iii)} & P_{0,x,\xi}(e=\infty) = 1 \quad \text{a.e. } (x,\xi) \in R^3\times R^3. \end{cases}$$

By using the temporal shift we can also show

$$(3.5)\qquad P_{s,x,\xi}(e=\infty) = 1 \quad \text{a.e. } (x,\xi) \in R^3\times R^3 \quad , \quad s \geqq 0.$$

This relation can be improved as follows.

Lemma 3.1 $P_{s,x,\xi}(e=\infty) = 1$ for every $(s,x,\xi) \in R_+\times R^3\times R^3$.

Proof. The strong Markov property of the family $\{P_{t,x,\xi}\}$ shows

$$\begin{aligned} &P_{s,x,\xi}(e=\infty) \\ &= P_{s,x,\xi}(\tau_2=\infty) + E_{s,x,\xi}[P_{s,x,\xi}(e=\infty\,|\,F_{\tau_2})\,;\,\tau_2<\infty] \\ &= P_{s,x,\xi}(\tau_2=\infty) + \int_{(s,\infty)\times R^3\times R^3} P_{t,\bar{x},\bar{\xi}}(e=\infty)\mu(dtd\bar{x}d\bar{\xi}). \end{aligned}$$

This implies the conclusion with the help of (3.5) since the measure $\mu(dtd\bar{x}d\bar{\xi}) = P_{s,x,\xi}(\tau_2\in dt, x_{\tau_2}\in d\bar{x}, \xi_{\tau_2}\in d\bar{\xi})$ is absolutely continuous with respect to the Lebesgue measure $dtd\bar{x}d\bar{\xi}$. □

The above observations are now summed up into the following.

Proposition 3.1 For each $(s,x,\xi) \in R_+\times R^3\times R^3$, there exists a unique probability measure $P = P_{s,x,\xi}$ on $D([s,\infty),R^3\times R^3)$ satisfying

$$(3.6)\quad \begin{cases} \text{(i)} \;\; P(y_s = (x,\xi)) = 1, \\ \text{(ii) for every } \psi \in C_0^\infty(R^3\times R^3), \\ \qquad \psi(y_t) - \int_s^t L_{f_1(r)}\psi(y_r)dr \quad , \quad t \geqq s, \\ \qquad \text{is a martingale relative to } (P,\{F_t\}). \end{cases}$$

Remark 3.1 By applying Varadhan's argument based on the results of Echeverria [3], we see the uniqueness of solutions to the weak forward equation (2.4), which is considered as an equation for nonnegative finite measures $f(t)$, when $\overline{f}_0$ is integrable.

4. Construction of processes.

We first discuss the case (I) where the initial data f_0 of the Boltzmann equation (1.1) is a probability density. We introduce the following assumptions.

[A.I.e] There exists a nonnegative solution $f_1(t,x,\xi)$, $0 \leqq t \leqq T$, to the weak version of the Boltzmann equation with initial data f_0:

$$\langle f_1(t),\psi\rangle = \langle f_0,\psi\rangle + \int_0^t \{\langle f_1(s),\xi\cdot\nabla_x\psi\rangle + \langle Q[f_1(s),f_1(s)],\psi\rangle\}ds,$$
$$0 \leqq t \leqq T \quad , \quad \psi \in C_0^\infty(R^3\times R^3), \quad (4.1)$$

and the function f_1 satisfies

$$\sup\{(1+|\xi|)\omega^{-1}(\xi)f_1(t,x,\xi)\,;\, t\in[0,T],\ x,\xi\in R^3\} < \infty. \quad (4.2)$$

[A.I.u] The Boltzmann equation (1.1) has a unique classical solution $f_1(t,x,\xi)$, $0 \leqq t \leqq T$, which satisfies the condition (4.2) and belongs to the class $C^1([0,T]\times R^3)$ in the variables (t,x) for each fixed $\xi \in R^3$.

Theorem 4.1 (i) Under the assumption [A.I.e] there exists a probability measure P on $D([0,T],R^3\times R^3)$ satisfying the condition [P.I].
(ii) Assume [A.I.u]. Then the martingale problem [P.I] has a unique solution satisfying $f(\cdot,\cdot,\xi) \in C^1([0,T]\times R^3)$, $\xi \in R^3$.

Proof. (i) We extend the domain of $f_1(t,x,\xi)$ to $R_+\times R^3\times R^3$ in such a way that it satisfies the condition in Section 2. Let P_{0,f_0} be the probability measure on $D([0,\infty),R^3\times R^3)$ defined by (3.3) with $\overline{f}_0 = f_0$. Then the density $f(t,x,\xi) = P_{0,f_0}(y_t\in dxd\xi)/dxd\xi$ which exists from (3.4)-(ii) solves the equation (2.4) with initial data f_0. However, since it holds the uniqueness of solutions to the equation (2.4) (see Remark 3.1), the condition [A.I.e] implies $f(t) = f_1(t)$, $t \in [0,T]$. Therefore the restriction P of P_{0,f_0} on $D([0,T],R^3\times R^3)$ satisfies the condition [P.I].
(ii) Since the solution $f(t,x,\xi)$ to the equation (4.1) satisfying $f(\cdot,\cdot,\xi) \in C^1([0,T]\times R^3)$, $\xi\in R^3$, is a classical solution to the Boltzmann equation, we get the conclusion from the uniqueness for the martingale problem (3.6). □

Remark 4.1 (i) The case of integrable initial datas was treated by Illner and Shinbrot [9], however, it is not clearly known that our assumption [A.I.e] or [A.I.u] holds exactly.
(ii) Our method works also in the case where the space domain (i.e., the domain of x) is bounded. In such a case, several solutions to the Boltzmann equation satisfying [A.I.e] or [A.I.u] are known; for example, see Ukai [19] or Shizuta [16] in the case of torus T^3(a cubic domain with periodic boundary condition) and Asano [1] or Shizuta and Asano [15] for other boundary conditions.

Next we discuss the case (II) where the initial data f_0 is not integrable. We need the following assumption.

[A.II] $\left\{\begin{array}{l}\text{There exists a nonnegative function } f_1(t,x,\xi),\ 0 \leq t \leq T, \text{ which} \\ \text{satisfies (4.2) and solves the Boltzmann equation with initial} \\ \text{data } f_0 \text{ in the sense that } u(t) = \omega^{-1}f_1(t) \text{ satisfies the} \\ \text{integral equation (2.3) with } u_0 = \omega^{-1}f_0.\end{array}\right.$

Theorem 4.2 Under the assumption [A.II] there exists an infinite system of stochastic processes $\{y_t^i = (x_t^i, \xi_t^i),\ t \in [0,T]\}_{i=1}^{\infty}$ satisfying the condition [P.II].

Proof. Take an increasing sequence $\{f_{0,N}(x,\xi)\}_{N=1}^{\infty}$ of nonnegative integrable functions which converge to $f_0(x,\xi)$ as N tends to ∞. We extend the domain of $f_1(t)$ to $[0,\infty)$ as in the proof of Theorem 4.1 and take the solution $u^{(N)}(t,x,\xi)$ of (2.3) with initial data $\omega^{-1}f_{0,N}$. Then (3.4) shows

$$P_{0,f_{0,N}}(y_t \in dxd\xi) = \omega(\xi)u^{(N)}(t,x,\xi)dxd\xi \quad , \quad t \geq 0,$$

where $P_{0,f_{0,N}}$ is a finite measure on $D([0,\infty),R^3\times R^3)$ defined by (3.3) with $\overline{f}_0 = f_{0,N}$. From (2.2) we see that $u^{(N)}(t,x,\xi)$ converges monotonically to the solution $u(t,x,\xi)$ to the equation (2.3) with initial data $\omega^{-1}f_0$. However Lemma 2.1-(i) and [A.II] show $\omega(\xi)u(t,x,\xi) = f_1(t,x,\xi)$ and therefore we have

$$(4.3) \qquad P_{0,f_0}(y_t \in dxd\xi) = f_1(t,x,\xi)dxd\xi \quad , \quad t \geq 0,$$

where P_{0,f_0} is a σ-finite measure on $D([0,\infty),R^3\times R^3)$ defined by (3.3). Now we determine the initial state $\{y_0^i=(x_0^i,\xi_0^i)\}_{i=1}^{\infty}$ of infinite particles as a point process on $R^3\times R^3$ with intensity $f_0(x,\xi)$ and move each particle independently according to the probability law P_{0,x_0^i,ξ_0^i}. Then (4.3) proves that the system of stochastic processes $\{y_t^i\}_{i=1}^{\infty}$ constructed in this way satisfies the desired condition [P.II]. □

Remark 4.2 We refer to Nishida and Imai [12], Ukai and Asano [20] and Asano [1] concerning the existence theorems for the Boltzmann equation with non-integrable initial data.

5. Euler limit.

Ukai and Asano [20] (see also Nishida [13]) gave a rigorous proof to the Euler limit for the Boltzmann equation with mean free path $\varepsilon > 0$:

$$(5.1)\quad \begin{cases} \dfrac{\partial f^\varepsilon}{\partial t} + \xi\cdot\nabla_x f^\varepsilon = \dfrac{1}{\varepsilon} Q[f^\varepsilon, f^\varepsilon] &, \quad (t,x,\xi) \in (0,T]\times R^3\times R^3, \\ f^\varepsilon(0) = f_0 &, \quad (x,\xi) \in R^3\times R^3. \end{cases}$$

Roughly speaking, they proved the following. If the initial data f_0 is sufficiently close to an absolute Maxwellian and analytic in x, then (a) the equation (5.1) has a unique classical solution $f^\varepsilon = f^\varepsilon(t,x,\xi)$ on a time interval $[0,T]$ independent of ε and (b) there exists a limit $f(t,x,\xi) = \lim_{\varepsilon\downarrow 0} f^\varepsilon(t,x,\xi)$ which is a local Maxwellian:

$$f(t,x,\xi) = \rho(t,x)\{2\pi T(t,x)\}^{-3/2}\exp\{-|\xi - v(t,x)|^2/2T(t,x)\},$$

whose hydrodynamical quantities (ρ,v,T) are unique solutions to the compressible Euler equation.

The aim of this section is to think their results over again from probabilistic point of view. From Theorem 4.2 combined with an estimate (5.5) below, we can construct an infinite system of stochastic processes $\{y_t^{i,\varepsilon} = (x_t^{i,\varepsilon}, \xi_t^{i,\varepsilon}),\ t \in [0,T]\}_{i=1}^\infty$ associated with the Boltzmann equation (5.1). For simplicity, we assume in the following the initial data $f_0 = f_0(x,\xi)$ is a local Maxwellian and the initial configuration $\{(x_0^i, \xi_0^i)\}_{i=1}^\infty$ of stochastic processes $\{(x_t^{i,\varepsilon}, \xi_t^{i,\varepsilon})\}_{i=1}^\infty$ is a Poisson point process on $R^3\times R^3$ with intensity $f_0(x,\xi)$. We shall prove that an $\mathcal{S}'(R^3)$-valued process

$$\mu_t^\varepsilon = \sum_{i=1}^\infty \delta_{x_t^{i,\varepsilon}} \quad , \quad 0 \leqq t \leqq T,$$

converges weakly to a process

$$(5.2)\quad \mu_t = \sum_{i=1}^\infty \delta_{x(t;x_0^i)} \ ,$$

where δ_x stands for the δ-distribution at x and $\{x(t;\cdot),\ 0 \leqq t \leqq T\}$ is a flow of diffeomorphisms on R^3 determined by the following ODE:

$$(5.3)\quad \begin{cases} \dfrac{dx}{dt}(t;x) = v(t,x(t;x)) \\ x(0;x) = x. \end{cases}$$

Before proving the weak convergence, we now summarize some estimates and properties of f^ε and f for later use, which can be derived from the results of Ukai and Asano.

(5.4) $\displaystyle\sup_{0\leq t\leq T, x,\xi} (1+|\xi|)^{\beta_0}|f^\varepsilon(t,x,\xi)-f(t,x,\xi)|\omega^{-1}(\xi) \xrightarrow[\varepsilon\downarrow 0]{} 0$, $\beta_0 > 5/2$,

(5.5) $\displaystyle\sup_{\varepsilon,t,x,\xi} f^\varepsilon(t,x,\xi)\omega^{-1}(\xi) < \infty$,

(5.6) $f(t,x,\xi)$ and therefore $\{\rho(t,x),v(t,x),T(t,x)\}$ are smooth in (t,x),

(5.7) $v \in C_b^{0,2}([0,T]\times R^3)$,

(5.8) $\displaystyle\inf_{t,x} \rho(t,x) > 0.$

We need a slightly more restrictive assumption on f_0 than in Ukai and Asano to get (5.8).

After Itô [10], using Hermite functions $\{h_n(t)\ , t \in R^1\}_{n=0}^\infty$, we introduce norms

$$\|f\|_p = \sum_{|n|=0}^{\infty} \prod_{i=1}^{3} (2n_i+1)^{2p}\langle f,h_n\rangle^2 \quad , \quad f \in S(R^3), \quad p \geqq 0,$$

where

$$n = (n_1,n_2,n_3) \in Z_+^3 \quad , \quad |n| = \sum_{i=1}^{3} n_i \ ,$$

$$h_n(x) = \prod_{i=1}^{3} h_{n_i}(x_i) \quad , \quad x = (x_1,x_2,x_3),$$

and $\langle\ ,\ \rangle$ stands for the inner product of the space $L^2(R^3,dx)$. As usual, we introduce a family of Hilbert spaces $S'_p(R^3)$, $p \geq 0$, with dual norms $\|\cdot\|_{-p}$, which are subspaces of $S'(R^3)$. Then the process $\{\mu_t^\varepsilon, t \in [0,T]\}$ induces a probability distribution Q^ε on the space $C([0,T],S'_p)$ with $p > 3$.

<u>Proposition 5.1</u> The family $\{Q^\varepsilon\ , \varepsilon > 0\}$ is tight.

<u>Proof</u>. It is enough to prove the following three equalities (see, e.g., Holley and Stroock [7]).

(5.9) $\displaystyle\lim_{M\to\infty} \sup_{\varepsilon} P(\|\mu_0^\varepsilon\|_{-p} > M) = 0,$

(5.10) $\displaystyle\lim_{n\to\infty} \sup_{\varepsilon} P(\sup_{0\leq t\leq T} \|\Pi_n^\perp\mu_t^\varepsilon\|_{-p} > \delta) = 0$ for every $\delta > 0$,

(5.11) $\displaystyle\lim_{\delta\downarrow 0} \sup_{\varepsilon} P(\sup_{\substack{|t-s|<\delta \\ t,s\in[0,T]}} \|\Pi_n\mu_t^\varepsilon-\Pi_n\mu_s^\varepsilon\|_{-p} > \alpha) = 0$ for every $n = 1,2,\cdots$ and $\alpha > 0$,

where

$$\Pi_n\mu = \sum_{|n|\leq n} {}_{S'}\langle\mu,h_n\rangle_S\, h_n$$

and $\Pi_n^\perp = I - \Pi_n$. However an equality

(5.12) $\quad <\mu_t^\varepsilon,\psi> = <\mu_s^\varepsilon,\psi> + \int_s^t <\eta_r^\varepsilon,\xi\cdot\nabla\psi(x)>dr \quad , \quad \psi \in \mathcal{S}(R^3),$

implies

$$E[\sup_{0\leq t\leq T} <\mu_t^\varepsilon,\psi>^2] \leq 2E[<\mu_0^\varepsilon,\psi>^2] + 2T\int_0^T E[<\eta_s^\varepsilon,\xi\cdot\nabla\psi>^2]ds,$$

where

$$\eta_s^\varepsilon = \sum_{i=1}^\infty \delta_{y_s^{i,\varepsilon}} \in \mathcal{S}'(R^3\times R^3).$$

Since η_s^ε is a Poisson random measure on $R^3\times R^3$ with intensity $f^\varepsilon(s,x,\xi)$, the estimate (5.5) shows

(5.13) $\quad E[<\eta_s^\varepsilon,\xi\cdot\nabla\psi>^2]$

$$= \int|\xi\cdot\nabla\psi|^2 f^\varepsilon(s,x,\xi)dxd\xi + \{\int\xi\cdot\nabla\psi\ f^\varepsilon(s,x,\xi)dxd\xi\}^2$$

$$\leq C_1\{\ \|\nabla\psi\|^2_{L^2(R^3)} + \|\nabla\psi\|^2_{L^1(R^3)}\}$$

$$\leq C_2\ \|\psi\|_q^2 \quad , \quad q > 5/2.$$

Therefore we get

$$E[\sup_{0\leq t\leq T} <\mu_t^\varepsilon,\psi>^2] \leq C_3\ \|\psi\|_q^2 \quad , \quad q > 5/2,$$

which implies (5.9) and (5.10). While (5.11) can be shown by

$$\lim_{\delta\downarrow 0}\sup_\varepsilon P(\sup_{\substack{|t-s|<\delta \\ t,s\in[0,T]}} |<\mu_t^\varepsilon,\psi> - <\mu_s^\varepsilon,\psi>| > \alpha) = 0 \quad , \quad \psi \in \mathcal{S}(R^3),$$

which follows from (5.12) and (5.13). □

Let Q be a probability distribution on the space $C([0,T],\mathcal{S}'_p)$ of the process $\mu_\cdot$ defined by (5.2). We use a usual perturbation technique to prove the weak convergence of Q^ε to Q. Noting that

$$\zeta(t,x,\xi) = \mathcal{B}^{-1}_{f(t,x,\cdot)}(\xi-v(t,x))$$

is a function in the space $\{L^2(R^3_\xi,f(t,x,\xi)d\xi)\}^3$ for every fixed $(t,x) \in [0,T]\times R^3$ (see Funaki [5]), we put

$$\psi^\varepsilon(t,x,\xi) = \psi(x) - \varepsilon\phi(x,\xi)\zeta(t,x,\xi)\cdot\nabla\psi(x)$$

for $\psi = \psi(x) \in C_0^\infty(R^3)$ and $\phi = \phi(x,\xi) \in C_0^\infty(R^3\times R^3)$. By a similar argument in Caflish [2], we see from (5.6) and (5.8) that

(5.14) $\quad \zeta(t,x,\xi)$ is smooth in (t,x) and locally bounded in ξ.

From an equality:

$$<\eta_t^\varepsilon,\psi^\varepsilon(t)> = <\eta_0^\varepsilon,\psi^\varepsilon(0)>$$

$$+ \int_0^t <\eta_s^\varepsilon,\{\frac{\partial}{\partial s} + \xi\cdot\nabla_x + \frac{1}{\varepsilon}\mathcal{B}_{f^\varepsilon(s,x,\cdot)}\}\psi^\varepsilon(s)>ds + \text{martingale},$$

we obtain

$$\langle\mu_t^\varepsilon,\psi\rangle - \langle\mu_0^\varepsilon,\psi\rangle - \int_0^t \langle\mu_s^\varepsilon, v(s,x)\cdot\nabla\psi\rangle ds = \sum_{i=1}^{5} \alpha_i^\varepsilon(t) - \alpha_i^\varepsilon(0),$$

where

$$\alpha_1^\varepsilon(t) = \varepsilon\langle\eta_t^\varepsilon, \phi(x,\xi)\zeta(t,x,\xi)\cdot\nabla\psi(x)\rangle,$$

$$\alpha_2^\varepsilon(t) = -\int_0^t \langle\eta_s^\varepsilon, \{\mathcal{B}_{f^\varepsilon(s,x,\cdot)} - \mathcal{B}_{f(s,x,\cdot)}\}\{\phi(x,\xi)\zeta(t,x,\xi)\cdot\nabla\psi(x)\}\rangle ds,$$

$$\alpha_3^\varepsilon(t) = -\varepsilon\int_0^t \langle\eta_s^\varepsilon, (\frac{\partial}{\partial s} + \xi\cdot\nabla_x)\{\phi(x,\xi)\zeta(t,x,\xi)\cdot\nabla\psi(x)\}\rangle ds,$$

$$\alpha_4^\varepsilon(t) = \int_0^t \langle\eta_s^\varepsilon, \mathcal{B}_{f(s,x,\cdot)}\{(1-\phi(x,\xi))\zeta(t,x,\xi)\cdot\nabla\psi(x)\}\rangle ds,$$

and $\alpha_5^\varepsilon(t)$ is a martingale with quadratic variational process

$$\langle\alpha_5^\varepsilon\rangle(t) = \int_0^t \langle\eta_s^\varepsilon, h^\varepsilon(s,x,\xi)\rangle ds,$$

$$h^\varepsilon(s,x,\xi) = \varepsilon\mathcal{B}_{f^\varepsilon(s,x,\cdot)}\{\phi(x,\xi)\zeta(t,x,\xi)\cdot\nabla\psi(x)\}^2$$
$$- 2\varepsilon\phi(x,\xi)\zeta(t,x,\xi)\cdot\nabla\psi(x)\mathcal{B}_{f^\varepsilon(s,x,\cdot)}\{\phi(x,\xi)\zeta(t,x,\xi)\cdot\nabla\psi(x)\}.$$

It is not difficult to prove

(5.15) $$\lim_{\varepsilon\downarrow 0} E[|\alpha_i^\varepsilon(t)|] = 0$$

for $i = 1,3,5$, by using (5.5) and (5.14). While, for $i = 2$, (5.15) is shown by (5.4) with the help of the following lemma (see Lemma 4.2 of Funaki [5]).

Lemma 5.1 There exists a positive constant $C = C_{\beta,\beta_0}$, $\beta \geqq 0$, $\beta_0 > 2$, such that

$$\|\Phi\|_\beta \leqq C\,\|\omega^{-1}\{f_1 - f_2\}\|_{\beta_0}\,\|\phi\|_{\beta+1}$$

holds for functions ϕ, f_1, f_2 of ξ, where

$$\Phi(\xi) = \omega(\xi)\{\mathcal{B}_{f_1} - \mathcal{B}_{f_2}\}\{\omega^{-1}(\xi)\phi(\xi)\}$$

and

$$\|\phi\|_\beta = \sup_\xi (1+|\xi|^\beta)|\phi(\xi)|.$$

As for the term $\alpha_4^\varepsilon(t)$ we have

(5.16) $$\sup_\varepsilon E[|\alpha_4^\varepsilon(t)|] \leqq C\int_0^t ds \int |(1-\phi(x,\xi))\zeta(s,x,\xi)\cdot\nabla\psi(x)|(1+|\xi|)\omega(\xi)dxd\xi.$$

Now we need the following assumption to control this integral.

[A.III] $|\zeta(s,x,\xi)| \leqq \text{const.}(1+|x|^p)\exp\{\alpha|\xi|^2/2\}$, $0 \leqq p < \infty$, $\alpha < 1/2$.

Note that this assumption holds at least for the Maxwellian molecules, i.e., in the case where $B(\theta,|\xi|)$ depends only on θ. The assumption

[A.III] implies that the right hand side of (5.16) can be taken arbitrarily small with suitable choice of the function $\phi \in C_0^\infty(R^3\times R^3)$ and therefore we get the following.

Lemma 5.2 Under the assumption [A.III] we have

$$\lim_{\varepsilon \downarrow 0} E^{Q^\varepsilon}[|\langle\mu_t,\psi\rangle - \langle\mu_s,\psi\rangle - \int_0^t \langle\mu_s, v(s,x)\cdot\nabla\psi\rangle ds|] = 0 \quad , \quad \psi \in C_0^\infty(R^3).$$

Now we can prove the following theorem.

Theorem 5.1 Assume [A.III], then Q^ε converges weakly to Q as ε tends to 0.

Proof. Since $\{Q^\varepsilon\}$ is tight, we may only prove that $Q^{\varepsilon'} \Rightarrow \bar{Q}$ (weakly as $\varepsilon' \downarrow 0$) implies $\bar{Q} = Q$. However Lemma 5.2 proves

$$\langle\mu_t,\psi\rangle = \langle\mu_0,\psi\rangle + \int_0^t \langle\mu_s, v(s,x)\cdot\nabla\psi\rangle ds \quad , \quad 0 \le t \le T, \ \psi \in \mathcal{D}, \quad \bar{Q}\text{-a.s.},$$

where $\mathcal{D}$ is a countable dense subset of $C_0^\infty(R^3)$. This with the help of (5.7) implies $\bar{Q} = Q$ and therefore we get the conclusion. □

Remark 5.1 The central limit theorem was discussed in Funaki [5] for the spatially homogeneous Boltzmann equation.

REFERENCES

[1] K. Asano, Local solutions to the initial and initial boundary value problem for the Boltzmann equation with an external force I, J. Math. Kyoto Univ. 24 (1984) 225-238.

[2] R.E. Caflish, The fluid dynamical limit of the nonlinear Boltzmann equation, Commun. Pure Appl. Math. 33 (1980) 651-666.

[3] P. Echeverria, A criterion for invariant measures of Markov processes, Z. Wahrscheinlichkeitstheorie verw. Gebiete 61 (1982) 1-16.

[4] T. Funaki, The diffusion approximation of the spatially homogeneous Boltzmann equation, Duke Math. J. 52 (1985) 1-23.

[5] ________ , The central limit theorem for spatially homogeneous Boltzmann equation, Technical Report #40, LRSP, Carleton Univ., Ottawa 1984.

[6] H. Grad, Asymptotic theory of the Boltzmann equation, II, in Rarefied gas dynamics I (edited by J.A. Laurmann) 25-59 (1963).

[7] R.A. Holley and D.W. Stroock, Generalized Ornstein-Uhlenbeck processes and infinite particle branching Brownian motions, Publ. RIMS Kyoto Univ. 14 (1978) 741-788.

[8] N. Ikeda and S. Watanabe, Stochastic differential equations and diffusion processes, North-Holland/Kodansha, Amsterdam/Tokyo 1981.

[9] R. Illner and M. Shinbrot, The Boltzmann equation: global existence for a rare gas in an infinite vacuum, Commun. Math. Phys. 95 (1984) 217-226.

[10] K. Itô, Foundations of stochastic differential equations in infinite dimensional spaces, Philadelphia, SIAM 1984.

[11] H.P. McKean, A class of Markov processes associated with nonlinear parabolic equations, Proc. Nat. Acad. Sci. 56 (1966) 1907-1911.

[12] T. Nishida and K. Imai, Global solutions to the initial value problem for the nonlinear Boltzmann equation, Publ. RIMS Kyoto Univ. 12 (1976) 229-239.

[13] T. Nishida, Fluid dynamical limit of the nonlinear Boltzmann equation to the level of the compressible Euler equation, Commun. Math. Phys. 61 (1978) 119-148.

[14] G.C. Papanicolaou, Asymptotic analysis of transport processes, Bull. Amer. Math. Soc. 81 (1975) 330-392.

[15] Y. Shizuta and K. Asano, Global solutions of the Boltzmann equation in a bounded convex domain, Proc. Japan Acad. 53 (1977) 3-5.

[16] Y. Shizuta, On the classical solution of the Boltzmann equation, Commun. Pure Appl. Math. 36 (1983) 705-754.

[17] A.-S. Sznitman, Equations de type Boltzmann, spatialement homogenes, Z. Wahrscheinlichkeitstheorie verw. Gebiete 66 (1984) 559-592.

[18] H. Tanaka, Probabilistic treatment of the Boltzmann equation of Maxwellian molecules, Z. Wahrscheinlichkeitstheorie verw. Gebiete 46 (1978) 67-105.

[19] S. Ukai, On the existence of global solutions of mixed problem for non-linear Boltzmann equation, Proc. Japan Acad. 50 (1974) 179-184.

[20] S. Ukai and K. Asano, The Euler limit and initial layer of the nonlinear Boltzmann equation, Hokkaido Math. J. 12 (1983) 311-332.

ISOTROPIC STOCHASTIC FLOWS AND A RELATED PROPERTY OF NON-RANDOM POTENTIAL FLOWS

T.E. Harris
Department of Mathematics, DRB 306
University of Southern California
Los Angeles, California 90089-1113

Summary

Sections 1 and 2 summarize some of the results of [2] and [8] about isotropic stochastic flows and their stability properties. In Sections 3 and 4, stability properties of isotropic flows and certain stirring processes are related to a mean shrinkage property of a randomly chosen small segment in deterministic potential flow with compact support in R^2 or R^3.

1. INTRODUCTION

A stochastic flow is a family of random mappings X_{st}, $0 \leq s \leq t < \infty$ of a space M into itself such that (a) X_{ss} = identity, $X_{tu} \circ X_{st} = X_{su}$, $0 \leq s \leq t \leq u < \infty$; and (b) $X_{s_1t_1}, X_{s_2t_2}, \ldots$ are independent if $s_1 \leq t_1 \leq s_2 \leq t_2 \leq \ldots$. Under the technical conditions commonly imposed, (b) may be replaced by (b'): the motion of each finite set of points is Markovian. We put $X_t = X_{0t}$, $t \geq 0$.

We shall restrict our attention to the case where $M = R^d$, $d \geq 2$ (for $d = 1$ see [5]), with spatial and temporal homogeneity. Moreover we consider only isotropic flows, where each finite-set process $(X_t(x_1),\ldots,X_t(x_k),\ t \geq 0)$ has the same law as $(G^TX_t(Gx_1),\ldots,G^TX_t(Gx_k),\ t \geq 0)$ whenever $G \in O(d)$. Isotropic flows have properties, such as the Markovian nature of two-point distances, which simplify their study. The results discussed below probably have analogues if only homogeneity is assumed.

Unless the contrary is stated, the finite-set motions are assumed to be diffusions. (The other case considered will be the "stirring processes" considered in Section 3.)

For an isotropic flow (of the diffusion type) the one-point motions are necessarily Brownian with drift 0. The process is determined by its generating covariance tensor $b^{pq}(x)$:

$$b^{pq}(x) = \lim_{t\downarrow 0} E \frac{(X_t^p(y+x) - y^p - x^p)(X_t^q(y) - y^q)}{t}, \quad p,q = 1,\ldots,d, \qquad (1.1)$$

where superscripts indicate coordinates. We assume b vanishes at ∞; otherwise a trivial adjustment is required. b is the covariance tensor of some mean-zero R^d-valued Gaussian field $(U(x),\ x \in R^d)$, which we may think of as the field of small displacements of the flow. The condition on b for isotropy is $b(Gx) = Gb(x)G^T$, $G \in O(d)$. We assume that b has continuous partial derivatives of order ≤ 4, ensuring the existence of a version of $(X_t(x),\ t \geq 0,\ x \in R^d)$ for which each X_t is a diffeomorphism of R^d onto R^d, with joint continuity in t and x. For further information on isotropic flows see LeJan [8] and Baxendale and Harris [2].

According to results of Ito [6] and Yaglom [12] the covariance tensor of an isotropic vector field, taking hereafter the normalization $b^{pq}(0) = \delta_{pq}$, may be written as a convex combination

$$b^{pq} = a_P b_P^{pq} + a_S b_S^{pq}, \qquad (1.2)$$

where b_P and b_S are normalized covariance tensors of, respectively, a potential (irrotational) field $U_P(x)$, i.e. $\frac{\partial U_P^p}{\partial x^q} = \frac{\partial U_P^q}{\partial x^p}$ and a solenoidal field $U_S(x)$, i.e. div $U_S(x) = 0$. A number of quantities connected with the corresponding flow, such as diffusion coefficients and the Lyapunov exponents, have corresponding convex representations.

We call an isotropic flow <u>potential</u> if $a_P = 1$ and <u>solenoidal</u> if $a_S = 1$. Solenoidal flows preserve Lebesgue measure; potential ones are characterized by a property analogous to zero curl, although the usual curl does not exist; see [2], Section 6.

2. Stability questions.

Without giving an exact definition of stability in a flow, we may think of it as a tendency toward shrinking or convergence. A weak criterion for stability would be

$$\lim_{|x_1-x_2| \to 0} P\{ \lim_{t\to\infty} |X_t(x_1) - X_t(x_2)| = 0\} = (?)\ 1. \tag{2.1}$$

Now (2.1) is never true for solenoidal isotropic flows; it is true for potential isotropic flows if $d = 2$ or 3 but not if $d \geq 4$. Hence we should expect some kind of stability for isotropic flows only if $d = 2$ or 3 and the potential component is sufficiently dominant.

We get the same picture from the tangent flow. Let T_x be the tangent space at some $x \in R^d$ and let v be a non-zero vector in T_x. Let $v_t = DX_t(x)v$ be the image of v under the differential map. Then log $|v_t|$ is a Wiener process with a drift μ_1 that is strictly negative iff $d = 2$ or 3 and the potential component is sufficiently dominant. See [2] or [8]. It is also known that if $v_{it} = DX_t(x)v_i$, $i = 1,2$, where $v_1, v_2 \in T_x$, $|v_i| \neq 0$, then v_{1t} and v_{2t} approach alignment in the same or opposite direction as $t \to \infty$. See [2] for discussion.

To relate this to the theory of Lyapunov exponents, we may, for present purposes, define the largest Lyapunov exponent by

$$\mu_1 = \lim_{t\to\infty} \frac{\log |v_{1t}|}{t}, \tag{2.2}$$

where μ_1 is the Wiener drift mentioned above. The limit for an isotropic flow is

$$\frac{1}{2}[(d-1)\beta_N - \beta_L] \tag{2.3}$$

where β_N, β_L are positive constants depending on the process; the ratio β_L/β_N lies between $(d-1)/(d+1)$ (for the solenoidal case) and 3 (for the potential case). It follows that $\mu_1 < 0$ iff $d = 2$ or 3 and the potential component is sufficiently dominant.

For more on Lyapunov exponents see [1], [3], and [8]. Carverhill's paper [3] gives a thorough discussion and many results concerning Lyapunov theory for stochastic flows.

On the basis of results proved in [3] for flows in compact manifolds, using results of Ruelle [11], one would expect the following statement S to hold for isotropic flows.

S. If $d = 2$ or 3 and $\mu_1 < 0$, then for each $\varepsilon > 0$ we can find $\delta > 0$ such that if B is a ball in R^d of diameter $< \delta$, then

$$P\{\lim_{t\to\infty} \operatorname{diam}(X_t(B)) = 0\} > 1 - \varepsilon.$$

As far as the author knows, S has not been proved for isotropic flows, although for $d = 2$ and $\beta_L > 5\,\beta_N/3$ rather than the hoped for condition $\beta_L > \beta_N$ corresponding to $\mu_1 < 0$, the required δ can be found for each ε. This follows from some properties of lengths of arcs carried by the flow, applied to the perimeter; see [2].

Results of a quite different nature about volumes are given in [8].

Although we shall not discuss arc lengths here, we note the related fact that certain configurations of tangent vectors originating in a single T_x tend to resist shrinkage of their total lengths even in cases where $\mu_1 < 0$. The effect is necessarily transient because of the tendency of tangent vectors to line up. Let $v_1,\dots,v_n \in T_x$, $v_{it} = DX_t(x)v_i$, $Z_t = \log \Sigma|v_{it}|$. From the Ito calculus we find $dZ_t = dY_t + a(t)dt$, where for the martingale Y we have $0 < c_1 \leq d\langle Y\rangle_t/dt \leq c_2$, and

$$a(t) = c_3 - c_4 \left(\sum_i |v_{it}|\right)^{-2} \sum_{i,j=1}^{n} \frac{(v_{it},v_{jt})^2}{|v_{it}||v_{jt}|}.$$

Here $c_3 > 0$ and $c_4 > 0$ are constants depending on the process. The double sum is $\leq (\Sigma|v_{it}|)^2$ (equality if all the v_{it} have one direction or its opposite) and can be shown to be $\geq (\Sigma|v_{it}|)^2/d$ (equality if $n = d$ and the v_{it} are mutually perpendicular with the same length.) In particular, if $d = 3$ and the flow is potential we find that $a(t) = -\frac{1}{2}\beta_N \equiv \mu_1$ in case the v_{it} are aligned and $a(t) = \beta_N/6$ in the perpendicular case.

This shows in a simplified form the reason for the dependence on shape of the length-process of a small arc, as given in [2]: A short arc is most resistant to shrinkage if it consists of d mutually perpendicular equal pieces.

3. Stability properties for randomly placed non-random potential flows.

The stability properties of isotropic potential stochastic flows in R^2 and R^3 noted in Section 2 raise the question whether in some average sense a smooth non-random potential flow in R^2 or R^3 is stable. One can give counter examples for flows that do not die away at ∞. For

simplicity, then, we consider only potential flows with compact support in R^d, $d \geq 2$. We shall see that a tangent vector, considered as the limiting case of a small segment, selected at a random point of the flow with a random orientation, has in a certain sense a tendency to shrink during a short time interval if $d = 2$ or 3.
Of course at any single point the flow may be either contractive or expansive.

Let $\psi : R^d \to R^1$ belong to C_0^4, with support in $B_K = \{x \in R^d : |x| < K\}$. Let $(u_\tau, \tau \in R^1)$ be the flow in R^d associated with the vector field grad ψ :

$$\frac{\partial u_\tau(x)}{\partial \tau} = \text{grad } \psi\, (u_\tau(x)), \qquad u_0(x) = x. \tag{3.1}$$

Since it is more convenient to randomize the flow than the tangent vector, define

$$\psi^{G,\alpha}(x) = \psi(G(x-\alpha)), \ \alpha \in B_K, \ G \in O(d). \tag{3.2}$$

The flow associated with $\psi^{G,\alpha}$ is then

$$u_\tau^{G,\alpha}(x) = \alpha + G^T u_\tau(G(x-\alpha)). \tag{3.3}$$

We shall give (α,G) the product probability measure having α uniform in B_K, while G has normalized Haar measure, denoted by dG.

Fix some $\tau > 0$. The differential of $u_\tau^{G,\alpha}$ at $x = 0$ is the random matrix $M_\tau = M_\tau^{G,\alpha}$,

$$M_\tau(i,j) = \frac{\partial u_\tau^{G,\alpha,i}(0)}{\partial x^j}, \tag{3.4}$$

the superscript i denoting the ith component. Since $u_\tau^{G,\alpha}$ is a diffeomorphism, M_τ is non-singular.

From the definitions we have

$$M_\tau = G^T N_\tau(-G\alpha)G, \tag{3.5}$$

where $N_\tau(y)$ is the matrix having $\partial_j u_\tau^i(y)$ in the ith row, j^{th} column. From the properties of Haar measure G is independent of $G\alpha$. It follows that M_τ is isotropic (i.e. for fixed

$H \in O(d)$, $H^T M_\tau H$ has the same law as M_τ).

Define $\mu_{1\tau} = E \log |M_\tau e|$, where e is an arbitrary unit vector. This may be considered as an intuitively reasonable measure of the shrinking or expanding tendency of M_τ. However, $\mu_{1\tau}$ is also the biggest Lyapunov exponent if independent copies of M_τ are multiplied. (See Newman [10].)

Theorem (3.6). $\mu_{1\tau}$ is given by

$$\frac{\tau^2(d-4)}{6(2\pi)^d} \int_{R^d} (\lambda_1)^4 \, H(\lambda)d\lambda + O(\tau^3),$$

where

$$H(\lambda) = \int_{R^d} |\hat{\psi}(G\lambda)|^2 dG \quad \text{and} \quad \hat{\psi}(\lambda) = \int_{R^d} e^{i(x,\lambda)}\psi(x)dx.$$

Except for scaling the coefficient of τ^2 is the largest Lyapunov exponent for the isotropic stochastic flow whose generating covariance tensor is given by (4.1).

Proof. By straightforward calculus

$$M_\tau(i,k) = \delta_{ik} + \tau(G^T D^2\psi(\beta)G_{ik}$$

$$+ \frac{1}{2}\tau^2\{G^T[D^2\psi(\beta)D^2\psi(\beta) + B(\beta)]G\}_{ik} + \theta_1 K_1 \tau^3, \quad \beta = -G\alpha. \tag{3.7}$$

Here $|\theta_1| < 1$, K_1 is a constant provided τ remains in a compact interval, and similarly for θ_2 and K_2 below; $B(\beta)$ is the $d \times d$ matrix with components

$$B_{jr}(\beta) = \sum_{s=1}^{d} \psi_{jrs}(\beta)\psi_s(\beta), \tag{3.8}$$

where $\psi_i = \partial\psi/\partial x_i$, $\psi_{ij} = \partial^2\psi/\partial x_i \partial x_j$, etc., and $D^2\psi(\beta)$ is the matrix $(\psi_{ij}(\beta))$. From the definition of $\mu_{1\tau}$, taking $e = (1,0,\ldots,0)$, we have

$$2\mu_{1\tau} = E \log (M_\tau(1,1)^2 + \cdots + M_\tau(d,1)^2) \tag{3.9}$$

$$= E \log \sum_{i=1}^{d} \{\delta_{i1} + \tau[G^T D^2\psi(\beta)G]_{i1}$$

$$+ \frac{1}{2}\tau^2[G^T(D^2\psi(\beta)D^2\psi(\beta) + B(\beta))G]_{i1} + \theta_2 K_2 \tau^3\}^2.$$

The right side of (3.9) is the expected value of $\log(1 + P\tau + Q\tau^2 + O(\tau^3)) = P\tau + (Q - \frac{1}{2}P^2)\tau^2 + O(\tau^3)$, where

$$P = 2[G^T D^2\psi(\beta)G]_{11}, \quad Q = \Sigma_1 + \Sigma_2,$$

$$\Sigma_1 = \sum_i [G^T D^2\psi(\beta)G]_{i1}^2 ,$$

$$\Sigma_2 = [G^T(D^2\psi(\beta)D^2\psi(\beta) + B(\beta))G]_{11}.$$

To take expectations, integrate over $O(d) \times B_K$ with respect to $dGd\beta/|B_K|$, $|B_K| = \text{vol}(B_K)$. Note that whenever the integrand contains a factor ψ_i, ψ_{ij}, etc., we may integrate instead over $O(d) \times R^d$. We find that $EP = 0$. Next, since

$$P^2 = 4 \sum_{jkrs} g_{j1}g_{k1}g_{r1}g_{s1}\psi_{jk}(\beta)\psi_{rs}(\beta),$$

where $G = (g_{ij})$, we have

$$\int_{R^d} P^2 d\beta = \frac{4}{(2\pi)^d} \sum_{j,k,r,s} g_{j1}g_{k1}g_{r1}g_{s1} \int_{R^d} \lambda_j\lambda_k\lambda_r\lambda_s|\hat{\psi}(\lambda)|^2 d\lambda \tag{3.10}$$
$$= \frac{4}{(2\pi)^d} \int (G^T\lambda, e_1)^4 \; |\hat{\psi}(\lambda)|^2 d\lambda ,$$

where e_1 is the unit vector whose first coordinate is 1 and we have used the relation

$$\int_{R^d} \psi_{jk}(\beta)\psi_{rs}(\beta)d\beta = (2\pi)^{-d} \int \lambda_j\lambda_k\lambda_r\lambda_s|\hat{\psi}(\lambda)|^2 d\lambda.$$

It follows that

$$EP^2 = 4(2\pi)^{-d} \; |B_k|^{-1} \int \lambda_1^4 \; H(\lambda)d\lambda, \tag{3.11}$$

$$H(\lambda) = \int_{O(d)} |\hat{\psi}(G\lambda)|^2 dG.$$

Similarly we find

$$E\;\Sigma_1 = (2\pi)^{-d} \; |B_K|^{-1} \int |\lambda|^2(\lambda_1)^2 H(\lambda)d\lambda, \tag{3.12}$$

$$E\;\Sigma_2 = 0.$$

From the spherical symmetry of H we have

$$\int \lambda_i^2 \lambda_j^2 H(\lambda)d\lambda = (1/3) \int \lambda_i^4 H(\lambda)d\lambda, \ i \neq j,$$

leading to (3.4). To complete the proof see (4.1) and the discussion following it.

4. Relation to stirring processes and other flows.

Stirring processes are special flows with jump-type trajectories. They have been studied by Lee [7] for discrete time, and were used in [4] as auxiliary processes in studying certain continuous flows; recently Matsumoto and Shigekawa [9] have widened the class of processes and have given general results on convergence to continuous flows on compact manifolds.

Here we shall only describe informally a particular stirring process which exhibits the connection between (3.6) and a class of (continuous) potential isotropic processes.

For $n = 1,2,\ldots$ let N_n be a Poisson point process (t_i,α_i,G_i) in $(0,\infty) \times R^d \times O(d)$ with mean density $n dt d\alpha dG$. Let $\mu_\tau^{G,\alpha}$ be as in Section 3 and let (X_t^n) be the flow (stirring process) constructed by applying $u_{1/\sqrt{n}}^{G_i \alpha_i}$ to R^d at time t_i. This makes sense even though the t_i are not well ordered; see the similar case in [4].

If we begin with a tangent vector $v \in T_x$ for some $x \in R^d$, we find that its successive values $v_1, v_2, \ldots$ after jumps under the flow X^n are the same as if we applied independent copies of the matrix M_τ of Section 3, with $\tau = 1/\sqrt{n}$. Moreover, considering the results in [9] for compact manifolds as well as a less comprehensive result for a simpler case in R^2 [4], we may conjecture that the sequence of process $(X_t^n, t > 0)$, whose values are C^1-diffeomorphisms $R^d \to R^d$, converges in process law to the (continuous) isotropic flow having the generating covariance tensor

$$b^{pq}(x) = \frac{1}{(2\pi)^d} \int_{R^d} e^{i(\lambda,x)} \lambda_p \lambda_q H(\lambda)d\lambda, \tag{4.1}$$

where H is as in Theorem (3.6). To achieve $b^{pq}(0) = \delta_{pq}$, we have taken

$$(2\pi)^d (\int (\lambda_1)^2 \, H(\lambda) d\lambda)^{-1} = 1. \qquad (4.2)$$

From [2] or [8], the largest Lyapunov exponent for this process is $\mu_1 = (d-4)(1/6)(2\pi)^{-d} \int \lambda_1^4 \, H(\lambda) d\lambda$, which except for the scaling factor is the same as in Theorem (3.6).

At least if d = 2, the class of isotropic potential flows having covariances of the form (4.1) is dense in an appropriate sense in the class of continuous isotropic potential flows having twice continuously differentiable covariance tensors. The argument is essentially the same as for the solenoidal case treated in [4]. In fact for this purpose it is sufficient to consider only potential functions ψ whose level lines are circles centered at the origin.

References

1. Baxendale, P. The Lyapunov spectrum of a stochastic flow of diffeomorphisms. Preprint, University of Aberdeen, Scotland.
2. Baxendale, P. and Harris, T.E. Isotropic stochastic flows. To appear, Ann. Prob. 1986.
3. Carverhill, A. Flows of stochastic dynamical systems: ergodic theory. Stochastics 14 273-318, 1985.
4. Harris, T. Brownian Motions on the homeomorphisms of the plane. Ann. Prob. 9 232-254, 1981.
7. Lee, W. Random stirring of the real line. Ann. Prob. 2, 1974, 580-592.
8. LeJan, Y. Exposants de Lyapunov pour les mourements Browniens isotropes. C.R. Paris Series I 299, 1984, 947-949. Also, a preprint, On isotropic Brownian motions, may now have been published.
9. Matsumoto, H. and Shigekawa, I. Limit theorems for stochastic flows of diffeomorphisms of jump type. Zeit. Wahr., 69, 1985, 507-540.
10. Newman, C. The distribution of Lyapunov exponents: exact results for random matrices. Comm. Math. Phys. To appear.
11. Ruelle, D. Ergodic theory of differentiable dynamical systems. Inst. des Hautes Etudes Scientific, Publications Math. No. 50, 1979, 275-306.
12. Yaglom, A. Some classes of random fields. Theory of Probability 2, 1957, (English translation by SIAM) 273-320.

EXPLOSION PROBLEMS FOR SYMMETRIC DIFFUSION PROCESSES

Kanji Ichihara

Department of Mathematics
College of General Education
Nagoya University
Nagoya 464, Japan

§1. Introduction.

Let (X_t, ζ, P_x), $x \in R^n$ be a diffusion process on R^n where ζ is the life time of X_t, i.e. $\lim_{t \uparrow \zeta(\omega)} X_t(\omega) = \infty$ if $\zeta(\omega) < +\infty$.

One of the basic problems for the process X_t is to find conditions for conservativeness and explosion. Such conditions for one dimensional processes have been established by Feller [1] in connection with the classification of boundary points. His conditions are given in terms of the scale and speed measures. In multidimensional cases, Hasminskii [3]([6]) has obtained sufficient conditions for conservativeness and explosion of diffusion processes which are constructed by means of Ito's stochastic differential equations.

We shall investigate these properties for symmetric diffusion processes associated with elliptic partial differential operators of self adjoint form with measurable coefficients. Hasminskii's idea is not applicable to our cases since the coefficients of the infinitesimal generator are not necessarily smooth. We make use of the theory of Dirichlet spaces instead to get conditions for conservativeness and explosion. General criteria for conservativeness and explosion are given in terms of the α-equilibrium potentials and α-capacities of the unit ball. These criteria are applied to obtain conditions on the coefficients for conservativeness and explosion. Some examples will be exhibited in the last section.

§2. Statement of Main theorems.

Let L be a strictly elliptic partial differential operator with measurable coefficients of the form:

$$L = \frac{1}{b} \sum_{i,j=1}^{n} \frac{\partial}{\partial x_i}\left(a_{ij} \frac{\partial}{\partial x_j}\right)$$

where $A(x) = (a_{ij}(x))_{i,j=1,..,n}$ is symmetric and $b > 0$. We assume that for each nonempty compact subset K of R^n, there exists a constant $\lambda = \lambda(K) > 1$ such that

$$\lambda^{-1}|\xi|^2 \leqq \sum_{i,j=1}^{n} a_{ij}(x)\xi_i\xi_j \leqq \lambda|\xi|^2$$

$$\lambda^{-1} \leqq b(x) \leqq \lambda$$

for all x in K and ξ in R^n. Define a nonnegative definite symmetric bilinear form $\mathcal{E}$ by

$$\mathcal{E}(u,v) = \int_{R^n} \sum_{i,j=1}^{n} \frac{\partial u}{\partial x_i} \frac{\partial v}{\partial x_j} a_{ij} dx$$

for $u,v \in C_0^\infty(R^n)$, the set of infinitely differentiable functions with compact support. $(\ ,\ )$ denotes the inner product in the real L^2-space $L^2(R^n,bdx)$. It is well known that under the above conditions on a_{ij}, b, the symmetric form $\mathcal{E}$ on $L^2(R^n,bdx)$ is closable. Consequently there exists an symmetric diffusion process except for a set of zero capacity whose Dirichlet form coincides with the smallest closed extension $\overline{\mathcal{E}}$ of $\mathcal{E}$. See Fukushima [2]. However, in the present case, we have a stronger result. In fact, there exists a unique minimal diffusion process (X_t,ζ,P_x), $x \in R^n$ whose resolvent R_λ^G restricted to a bounded open set G is continuous in the following sense; for any $f \in L^p(G,bdx)$ with $p > n$, $R_\lambda^G f(x)$ is continuous on $\overline{G}$. See Tomisaki [8]. Note that the infinitesimal generator of the diffusion coincides with the operator L.

Let B_n denote the closed unit ball centered at the origin. Let us introduce the first hitting time τ_0 of B_n by X_t, i.e.

$$\tau_0 = \begin{cases} \inf\{t \geqq 0 \mid |X_t| \leqq 1\} & \text{if such } t \text{ exists} \\ +\infty & \text{otherwise.} \end{cases}$$

The α-equilibrium potential $e_\alpha(x)$ is defined by

$$e_\alpha(x) = \begin{cases} E_x[e^{-\alpha\tau_0}] & , \quad \alpha > 0 \\ P_x[\tau_0 < \zeta] & , \quad \alpha = 0 \end{cases}$$

for $x \in R^n$. Set $\mathcal{E}_\alpha(u,v) = \mathcal{E}(u,v) + \alpha(u,v)$, $\alpha \geqq 0$ and denote by $\mathcal{F}_\alpha = \mathcal{F}_\alpha(R^n)$ the closure of $C_0^\infty(R^n)$ with respect to the norm $\mathcal{E}_\alpha$. It is well known that $(\mathcal{F}_\alpha, \mathcal{E}_\alpha)$, $\alpha > 0$ is a Hilbert space and that $(\mathcal{F}_0, \mathcal{E}_0)$ is a Hilbert space if and only if X_t is transient. (See Fukushima [2]). The α-capacity $C_\alpha(B_n)$ of the unit ball B_n is defined as follows. For $\alpha > 0$,

$$C_\alpha(B_n) = \inf_{\substack{u \in \mathcal{F}_\alpha \\ u \geq 1,\ dx\text{-a.e. on } B_n}} \mathcal{E}_\alpha(u,u)$$

and for $\alpha = 0$ and X_t being transient,

$$C_0(B_n) = \inf_{\substack{u \in \mathcal{F}_0 \\ u \geq 1,\ dx\text{-a.e. on } B_n}} \mathcal{E}_0(u,u).$$

Lemma 1 (Fukushima [2]). For each $\alpha > 0$, $e_\alpha \in \mathcal{F}_\alpha$ and

$$C_\alpha(B_n) = \mathcal{E}_\alpha(e_\alpha, e_\alpha).$$

Moreover, if X_t is transient, $e_0 \in \mathcal{F}_0$ and

$$C_0(B_n) = \mathcal{E}_0(e_0, e_0).$$

Remark 1. A simple variational technique shows that under the same conditions as in Lemma 1,

$$\mathcal{E}_\alpha(e_\alpha, u) \geq 0$$

for any $u \in \mathcal{F}_\alpha$ with $u \geq 0$, dx - a.e. on B_n.

We are now in a position to state fundamental criteria for conservativeness and explosion of the diffusion process X_t.

Theorem 1. The following conditions are equivalent.

(i) X_t is conservative, i.e. $P.[\zeta = +\infty] = 1$ on R^n.

(ii) For some $\alpha > 0$,

$$\int_{R^n} \alpha e_\alpha(x) b(x)\, dx = C_\alpha(B_n)$$

(iii) For all $\alpha > 0$,

$$\int_{R^n} \alpha e_\alpha(x) b(x) dx = C_\alpha(B_n).$$

Theorem 2. Suppose X_t is transient. Then the limit of $\int_{R^n} \alpha e_\alpha(x) b(x) dx$ as $\alpha \downarrow 0$ exists and

$$\lim_{\alpha \downarrow 0} \int_{R^n} \alpha e_\alpha(x) b(x) dx \begin{cases} = C_0(B_n) & , \text{ when } P.[\zeta = +\infty] = 1 \text{ on } R^n. \\ \in (0, C_0(B_n)), & \text{ when } 0 < P.[\zeta = +\infty] < 1 \text{ on } R^n. \\ = 0 & , \text{ when } P.[\zeta = +\infty] = 0 \text{ on } R^n. \end{cases}$$

Corollary 1. Suppose X_t is transient. If $\int_{R^n} e_0(x) b(x) dx$ is finite, then explosion of X_t is sure, i.e. $P.[\zeta < +\infty] = 1$ on R^n.

Corollary 2. Suppose X_t is transient. If the total mass of the speed measure is finite, i.e. $b \in L^1(R^n, dx)$, then explosion of X_t is sure.

In order to give conditions on coefficients a_{ij}, b for conservativeness, we define

$$A_+(r) = \int_{S^{n-1}} \langle A(r\sigma)\sigma, \sigma \rangle d\sigma \quad , \quad r > 0$$

$$B_+(r) = \sup_{\sigma \in S^{n-1}} \left\{ \frac{\langle A(r\sigma)\sigma, \sigma \rangle}{b(r\sigma)} \right\} , \quad r > 0$$

where $\langle \, , \, \rangle$ denotes the inner product in R^n and $d\sigma$ is the uniform measure on the unit sphere S^{n-1}. The following theorem gives a sufficient condition for conservativeness.

Theorem 3. If for some $\alpha > 0$,

$$\lim_{r \uparrow +\infty} \frac{\exp\left\{ -\int_1^r \frac{2\sqrt{\alpha}}{\sqrt{B_+(s)}} ds \right\}}{\int_r^{+\infty} s^{1-n} A_+(s)^{-1} ds} = 0,$$

then X_t is conservative.

It should be remarked that X_t is recurrent if the denominator in the assumption of Theorem 3 is infinite. See Ichihara [4].

As for explosion of X_t, we have

Theorem 4. If

$$\int_{S^{n-1}} d\sigma \int_1^{+\infty} r^{1-n} \langle A(r\sigma)^{-1}\sigma,\sigma\rangle \left[\int_1^r s^{n-1} b(s\sigma)ds\right]^2 dr < +\infty,$$

then explosion of X_t is sure. Here $A(x)^{-1}$ is the inverse matrix of $A(x)$.

§3. Proof of Theorems.

In the proofs of the fundamental criteria (Theorems 1 and 2), the following lemma plays an important role.

Lemma 2 (Fukushima [2]). (i) For each $\alpha > 0$, there exists a unique finite measure μ_α with its support in B_n such that

$$\mathcal{E}_\alpha(e_\alpha, u) = \int_{B_n} \tilde{u}(x)\mu_\alpha(dx) \quad , \quad u \in \mathcal{F}_\alpha .$$

(ii) For $\alpha = 0$ and X_t being transient, there exists a unique finite measure μ_0 with its support in ∂B_n, the boundary of B_n, such that

$$\mathcal{E}_0(e_0, u) = \int_{\partial B_n} \tilde{u}(x)\mu_0(dx) \quad , \quad u \in \mathcal{F}_0 .$$

Here $\tilde{u}$ denotes any quasi-continuous modification of u.

The above measure μ_α is called the α-equilibrium measure of B_n. The total mass of μ_α is equal to the α-capacity $C_\alpha(B_n)$, i.e.

$$\mu_\alpha(B_n) = C_\alpha(B_n) \quad , \quad \alpha > 0$$

and

$$\mu_0(\partial B_n) = C_0(B_n) \quad .$$

Proof of Theorem 1. Let f be an infinitely differentiable function with compact support. Define the α-order Green measure $G_\alpha(x,dy)$ and Green operator G_α by

$$G_\alpha(x,dy) = \int_0^{+\infty} e^{-\alpha t} P_x[X_t \in dy, t < \zeta]dt$$

and

$$G_\alpha f(x) = \int_{R^n} G_\alpha(x,dy)f(y).$$

Then
$$\int_{R^n} \alpha e_\alpha(x) f(x) b(x) dx = \alpha(e_\alpha, f) = \alpha \mathcal{E}_\alpha(e_\alpha, G_\alpha f).$$

Applying Lemma 2 to $u = G_\alpha f$, we obtain

$$\int_{R^n} \alpha e_\alpha(x) f(x) b(x) dx = \alpha \int_{B_n} G_\alpha f(x) \mu_\alpha(dx).$$

Letting $f \uparrow 1$, it follows that

$$\int_{R^n} \alpha e_\alpha(x) b(x) dx = \alpha \int_{B_n} G_\alpha 1(x) \mu_\alpha(dx).$$

Since
$$G_\alpha 1(x) = \int_0^\infty e^{-\alpha t} E_x[1 ; t < \zeta] dt = E_x[\int_0^\zeta e^{-\alpha t} dt] = \frac{1}{\alpha}\{1 - E_x[e^{-\alpha\zeta}]\},$$

we obtain

$$\int_{R^n} \alpha e_\alpha(x) b(x) dx = C_\alpha(B_n) - \int_{B_n} E_x[e^{-\alpha\zeta}] \mu_\alpha(dx).$$

Now the implication (i) $\Longrightarrow$ (ii) and (iii) is obvious. Conversely, we assume that (ii) holds for some $\alpha > 0$. Then $\int_{B_n} E_x[e^{-\alpha\zeta}] \mu_\alpha(dx) = 0$.

Consequently, $E_x[e^{-\alpha\zeta}] = 0$, μ_α - a.e. on B_n. Thus $P_x[\zeta = +\infty] = 1$, μ_α - a.e. on B_n. Fix a point $x_0 \in B_n$ such that

(1) $$P_{x_0}[\zeta = +\infty] = 1.$$

Set $$q_0(x) = P_x[\zeta = +\infty] \quad \text{for} \quad x \in R^n.$$

The Markov property implies that

(2) $$q_0(x) = P_x[\zeta = +\infty] = P_x[t < \zeta, \zeta = +\infty] = E_x[P_{X_t}[\zeta = +\infty] ; t < \zeta] = E_x[q_0(X_t), t < \zeta].$$

Taking the Laplace transform of both sides of (2), we obtain

(3) $$\frac{1}{\alpha} q_0(x) = G_\alpha q_0(x).$$

Then (1) together with (3) imply that $q_0(x) = 1$, $G_\alpha(x_0, \cdot)$ - a.e. on R^n. On the other hand, Theorem 9.2 in Tomisaki [8] tells us that $G_\alpha(x, \cdot)$, $x \in R^n$ are equivalent to each other. Thus we have for all $x \in R^n$.

$$q_0(x) = \alpha G_\alpha q_0(x) = \alpha G_\alpha 1(x).$$

Hence

$$P_x[\zeta = +\infty] = 1 - E_x[e^{-\alpha\zeta}]$$

for all $x \in R^n$, which gives $P.[\zeta = +\infty] = 1$ on R^n. q.e.d.

For the proof of Theorem 2, we need the following lemmas.

Lemma 3. Suppose X_t is transient. Then $\lim_{\alpha \downarrow 0} \mathcal{E}_0(e_\alpha - e_0, e_\alpha - e_0) = 0$. In particular, we have

$$\lim_{\alpha \downarrow 0} C_\alpha(B_n) = C_0(B_n).$$

This lemma is proved by a simple argument based on the Dirichlet principle. The next lemma is easily derived from Lemmas 2 and 3.

Lemma 4. Suppose X_t is transient. Then it holds that for any continuous function $f(x)$ on R^n,

$$\lim_{\alpha \downarrow 0} \int_{B_n} f(x)\mu_\alpha(dx) = \int_{\partial B_n} f(x)\mu_0(dx).$$

Proof of Theorem 2. As was shown in the proof of Theorem 1, we have

$$\int_{R^n} \alpha e_\alpha(x)b(x)dx = \int_{B_n} \alpha G_\alpha 1(x)\mu_\alpha(dx).$$

Note that $\alpha G_\alpha 1(x) = 1 - E_x[e^{-\alpha\zeta}]$ is monotone nonincreasing and is convergent to $P_x[\zeta = +\infty] = 1 - P_x[\zeta < +\infty]$ as $\alpha \downarrow 0$. Besides, we can prove that $E_x[e^{-\alpha\zeta}]$, $\alpha > 0$ and $P_x[\zeta < +\infty]$ are continuous in $x \in R^n$. (See Remark 3.1, Ichihara [5] for the details.) Consequently the convergence $\alpha G_\alpha 1(x) \downarrow q_0(x) = P_x[\zeta = +\infty]$ as $\alpha \downarrow 0$ is uniform on each compact subset of R^n. This together with Lemma 4 implies

$$\lim_{\alpha \downarrow 0} \int_{R^n} \alpha e_\alpha(x)b(x)dx = \lim_{\alpha \downarrow 0} \int_{B_n} \alpha G_\alpha 1(x)\mu_\alpha(dx) = \int_{\partial B_n} q_0(x)\mu_0(dx).$$

Suppose now that

$$\lim_{\alpha \downarrow 0} \int_{R^n} \alpha e_\alpha(x)b(x)dx = C_0(B_n).$$

Then we have

$$\int_{\partial B_n} q_0(x)\mu_0(dx) = C_0(B_n).$$

Since the total measure of μ_0 is the capacity $C_0(B_n)$, this equality implies $q_0(x) = 1$, μ_0- a.e. on ∂B_n. Thus the same argument as in the proof of Theorem 1 shows that $q_0(x) = P_x[\zeta = +\infty]$ is identically equal to 1 on R^n.

We next consider the case

$$\lim_{\alpha \downarrow 0} \int_{R^n} \alpha e_\alpha(x) b(x) dx = 0.$$

Then it turns out to be

$$\int_{\partial B_n} q_0(x)\mu_0(dx) = 0,$$

which is equivalent to $q_0(x) = 0$, μ_0 - a.e. on ∂B_n. Again, by the same argument as in the first case, we can verify $q_0(x) = 0$ on R^n.

The rest follows from the proofs of the first two cases. q.e.d.

For the proof of Theorem 3, we need the following lemmas.

A function u is said to belong to $H^{1,2}_{loc}(R^n)$ if $u \in L^2_{loc}(R^n)$ and if its distribution derivatives $\frac{\partial u}{\partial x_i} \in L^2_{loc}(R^n)$.

<u>Lemma 5</u>. Let $\mathbf{f} = (f_i)_{i=1,\dots,n}$ where $f_i \in H^{1,2}_{loc}(R^n)$.

Define
$$\operatorname{div} \mathbf{f} = \sum_{i=1}^{n} \frac{\partial f_i}{\partial x_i}$$

and

$$F(r) = \int_{|x|=r} \langle \mathbf{f}, \mathbf{n} \rangle dS.$$

Then the derivative of $F(r)$ in the distribution sense is given by $\frac{d}{dr} F(r) = \int_{|x|=r} \operatorname{div} \mathbf{f}\, dS$ for a.a. $r > 0$.

Here $\mathbf{n} = \frac{x}{|x|}$ and dS is the surface element on $|x| = r$.

<u>Lemma 6</u>. Suppose that $u \in H^{1,2}_{loc}(R^n)$ and that $\mathbf{v} = (v_i)_{i=1,\dots,n}$, $v_i \in H^{1,2}_{loc}(R^n)$.

Then

$$\operatorname{div}(u\,\mathbf{v}) = \langle \nabla u, \mathbf{v} \rangle + u \operatorname{div} \mathbf{v},$$

where

$$\nabla u \quad \text{denotes} \quad \left(\frac{\partial u}{\partial x_i}\right)_{i=1,\dots,n}.$$

Since the function e_α is a weak solution of the Dirichlet problem:

$$\begin{cases} \frac{1}{b} \sum_{i,j=1}^{n} \frac{\partial}{\partial x_i}(a_{ij} \frac{\partial}{\partial x_j} u) = \alpha u & \text{in} \quad |x| > 1 \\ u(x) = 1 & \text{on} \quad |x| = 1, \end{cases}$$

by taking the continuous modification if necessary, we may assume that e_α is locally, uniformly Hölder continuous in $|x| > 1$. (See Remark 3.1, Ichihara [5].) Furthermore, without loss of generality, we may assume that a_{ij} and b are infinitely differentiable in a neighbourhood of S^{n-1}, because conservativeness and explosion depend only on the behaviors of the coefficients near infinity. Under the assumption, e_α is proved to be twice continuously differentiable on $1 \leqq |x| \leqq 1 + \varepsilon_0$ with a positive constant ε_0. (See e.g. Mizohata [7].) We have

$$C_\alpha(B_n) = \int_{R^n} \{<A\nabla e_\alpha, \nabla e_\alpha> + \alpha e_\alpha^2 b\} dx ,$$

$$= \int_{B_n} \alpha b dx + \int_{|x|>1} \{<A\nabla e_\alpha, \nabla e_\alpha> + \alpha e_\alpha^2 b\} dx ,$$

by the Dirichlet principle,

$$= \int_{B_n} \alpha b dx + \int_{|x|>1} \{<A\nabla e_\alpha, \nabla u> + \alpha e_\alpha u b\} dx$$

for $u \in \mathcal{F}_\alpha$ with $u = 1$ on B_n. Setting

$$u = u_\varepsilon = \begin{cases} 1 & , \text{ for } |x| < 1 \\ 1 - \frac{1}{\varepsilon}(|x| - 1), & \text{for } 1 < |x| < 1 + \varepsilon \\ 0 & , \text{ otherwise} \end{cases}$$

for $\varepsilon > 0$, we have

$$C_\alpha(B_n) = \int_{B_n} \alpha b dx - \frac{1}{\varepsilon} \int_{1<|x|<1+\varepsilon} <A\nabla e_\alpha, \mathbf{n}> dx + \int_{1<|x|<1+\varepsilon} \alpha e_\alpha u_\varepsilon b dx.$$

Letting $\varepsilon \downarrow 0$, we obtain

$$C_\alpha(B_n) = \int_{B_n} \alpha b dx - \int_{|x|=1} \langle A\nabla e_\alpha, \mathbf{n}\rangle dS.$$

Applying Lemma 5 to $\mathbf{f} = A\nabla e_\alpha$, we obtain

$$\int_{1<|x|<r} \alpha e_\alpha b dx = \int_{1<|x|<r} \operatorname{div}(A\nabla e_\alpha) dx = \int_1^r du \int_{|x|=u} \operatorname{div}(A\nabla e_\alpha) dS$$
$$= \int_1^r du \frac{d}{du} \int_{|x|=u} \langle A\nabla e_\alpha, \mathbf{n}\rangle dS$$
$$= \int_{|x|=r} \langle Ae_\alpha, \mathbf{n}\rangle dS - \int_{|x|=1} \langle A\nabla e_\alpha, \mathbf{n}\rangle dS$$

for a.a. $r > 1$. Hence

$$\int_{|x|=r} \langle A\nabla e_\alpha, \mathbf{n}\rangle dS \quad , \quad \text{a.a. } r > 1$$

is monotone nondecreasing as $r \uparrow +\infty$. Furthermore, it is nonpositive for a.a. $r > 1$. Indeed, as was shown in the proof of Theorem 1,

$$C_\alpha(B_n) \geqq \int_{R^n} \alpha e_\alpha b dx \geqq \int_{|x|<r} \alpha e_\alpha b dx \quad \text{for} \quad r > 1.$$

Since $C_\alpha(B_n) = \int_{B_n} \alpha b dx - \int_{|x|=1} \langle A\nabla e_\alpha, \mathbf{n}\rangle dS$ and

since $\int_{|x|<r} \alpha e_\alpha b dx = \int_{B_n} \alpha b dx + \int_{|x|=r} \langle A\nabla e_\alpha, \mathbf{n}\rangle dS - \int_{|x|=1} \langle A\nabla e_\alpha, \mathbf{n}\rangle dS$,

we obtain $\int_{|x|=r} \langle A\nabla e_\alpha, \mathbf{n}\rangle dS \leqq 0$ for a.a. $r > 1$.

Define

$$a_\alpha = \lim_{\text{a.a.}\, r \to +\infty} \int_{|x|=r} \langle A\nabla e_\alpha, \mathbf{n}\rangle dS.$$

Taking the limit as $r \uparrow +\infty$, we obtain

$$a_\alpha - \int_{|x|=1} \langle A\nabla e_\alpha, \mathbf{n}\rangle dS + \int_{B_n} \alpha b dx = \int_{R^n} \alpha b e_\alpha dx$$
$$\leqq C_\alpha(B_n)$$
$$= - \int_{|x|=1} \langle A\nabla e_\alpha, \mathbf{n}\rangle dS + \int_{B_n} \alpha b dx.$$

Thus we can arrive at the following interesting criterion for conservativeness.

Proposition 1. X_t is conservative if and only if $a_\alpha = 0$.

Proof of Theorem 3. It suffices to verify $a_\alpha = 0$ under the hypothesis.

Let ϕ be an infinitely differentiable function with compact support. Furthermore, we assume that ϕ is rotationally symmetric about the origin.

$$\int_{R^n} \operatorname{div}(e_\alpha A\nabla e_\alpha)\phi dx = -\int_{R^n} e_\alpha \langle A\nabla e_\alpha, \mathfrak{n}\rangle \phi'(|x|)dx.$$

By Lemma 6, the left-hand side is equal to

$$\int_{R^n} \langle A\nabla e_\alpha, \nabla e_\alpha\rangle \phi dx + \int_{R^n} e_\alpha \operatorname{div}(A\nabla e_\alpha)\phi dx = \int_{R^n} \{\langle A\nabla e_\alpha, \nabla e_\alpha\rangle + \alpha e_\alpha^2 b\}\phi dx.$$

Since ϕ is a radial function, we have the following equality

$$\int_0^\infty \phi(r)dr \int_{|x|=r} \{\langle A\nabla e_\alpha, \nabla e_\alpha\rangle + \alpha e_\alpha^2 b\}dS = -\int_0^{+\infty} \phi'(r)dr \int_{|x|=r} e_\alpha \langle A\nabla e_\alpha, \mathfrak{n}\rangle dS,$$

which gives

$$\int_{|x|=r} e_\alpha \langle A\nabla e_\alpha, \mathfrak{n}\rangle dS - \int_{|x|=1} \langle A\nabla e_\alpha, \mathfrak{n}\rangle dS = \int_{1<|x|<r} \{\langle A\nabla e_\alpha, \nabla e_\alpha\rangle + \alpha e_\alpha^2 b\}dx$$

for a.a. $r > 1$. Consequently,

$$V_\alpha(r) \overset{\text{def.}}{=} \int_{|x|>r} \{\langle A\nabla e_\alpha, \nabla e_\alpha\rangle + \alpha e_\alpha^2 b\}dx = -\int_{|x|=r} e_\alpha \langle A\nabla e_\alpha, \mathfrak{n}\rangle dS$$

for a.a. $r > 1$. We now compute the upper bound for $V_\alpha(r)$.

$$\begin{aligned}
V_\alpha(r) &\leqq \int_{|x|=r} \sqrt{\langle A\nabla e_\alpha, \nabla e_\alpha\rangle}\,\sqrt{\langle A\mathfrak{n}, \mathfrak{n}\rangle}\, e_\alpha dS \\
&= \int_{|x|=r} 2\sqrt{\langle A\nabla e_\alpha, \nabla e_\alpha\rangle}\,\sqrt{\alpha e_\alpha^2 b}dS\, \frac{1}{2\sqrt{\alpha}}\sqrt{B_+(r)} \\
&\leqq \frac{\sqrt{B_+(r)}}{2\sqrt{\alpha}} \int_{|x|=r} \{\langle A\nabla e_\alpha, \nabla e_\alpha\rangle + \alpha e_\alpha^2 b\}dS \\
&= -\frac{\sqrt{B_+(r)}}{2\sqrt{\alpha}} \frac{dV_\alpha(r)}{dr} \qquad \text{for a.a. } r > 1.
\end{aligned}$$

Solving this differential inequality, we obtain

$$V_\alpha(r) \leqq V_\alpha(1)\exp\{-\int_1^r \frac{2\sqrt{\alpha}}{\sqrt{B_+(s)}}\}ds\} \qquad \text{for all} \quad r > 1.$$

As for the lower bound, we have

$$V_\alpha(r) \geq \int_{|x|>r} <A\nabla e_\alpha, \nabla e_\alpha> dx = \int_r^{+\infty} ds \int_{|x|=s} <A\nabla e_\alpha, \nabla e_\alpha> dS.$$

The Schwarz inequality gives

$$\int_{|x|=s} <A\nabla e_\alpha, \nabla e_\alpha> dS \int_{|x|=s} <A\mathfrak{n}, \mathfrak{n}> dS \geq [\int_{|x|=s} <A\nabla e_\alpha, \mathfrak{n}> dS]^2.$$

Combining these, we obtain

$$V_\alpha(1)\exp\{-\int_1^r \frac{2\sqrt{\alpha}}{\sqrt{B_+(s)}} ds\} \geq \int_r^{+\infty} [\int_{|x|=s} <A\nabla e_\alpha, \mathfrak{n}> dS]^2 s^{1-n} A_+(s)^{-1} ds.$$

By the previous argument on a_α, we see that

$$|a_\alpha| \leq |\int_{|x|=s} <A\nabla e_\alpha, \mathfrak{n}> dS| \quad \text{for} \quad \text{a.a. } s > 1.$$

Thus we have

$$|a_\alpha|^2 \leq \frac{V_\alpha(1)\exp\{-\int_1^r \frac{2\sqrt{\alpha}}{\sqrt{B_+(s)}} ds\}}{\int_r^{+\infty} s^{1-n} A_+(s)^{-1} ds},$$

which tends to zero as $r \uparrow +\infty$ by the assumption. q.e.d.

Proof of Theorem 4. We first note that according to Ichihara [4], X_t is transient under the assumption in Theorem 4. In order to prove Theorem 4, it suffices to verify the condition in Corollary 1. Introduce

$$\sigma_\rho = \begin{cases} \inf\{t \geq 0 \mid |X_t| \geq \rho\} & \text{if such } t \text{ exists} \\ +\infty & \text{otherwise} \end{cases}$$

for $\rho > 1$. Define $e_0^\rho(x) = P_x[\tau_0 < \sigma_\rho]$. It is obvious that $e_0^\rho(x)$ is monotone nondecreasing and is convergent to $e_0(x)$ as $\rho \to +\infty$. Thus we have

$$\lim_{\rho\uparrow+\infty} \int_{1<|x|<\rho} e_0^\rho(x)b(x)dx = \int_{|x|>1} e_0(x)b(x)dx$$

by virtue of the monotone increasing theorem. We now estimate, by polar coordinates,

$$\int_{\rho>|x|>1} e_0^\rho(x)b(x)dx = \int_{S^{n-1}} d\sigma \int_1^\rho r^{n-1} b(r\sigma) e_0^\rho(r\sigma) dr$$

$$= \int_{S^{n-1}} d\sigma \int_1^\rho r^{n-1} b(r\sigma)\,dr \int_r^\rho - \frac{\partial e_0^\rho}{\partial s}(s\sigma)\,ds$$

$$\leqq \int_{S^{n-1}} d\sigma \int_1^\rho \left| \frac{\partial e_0^\rho(s\sigma)}{\partial s} \right| ds \int_1^s r^{n-1} b(r\sigma)\,dr$$

$$= \int_{S^{n-1}} d\sigma \int_1^\rho |<\sqrt{A}\nabla e_0^\rho, \sqrt{A}^{-1}\mathfrak{n}>(s\sigma)|\,ds \int_1^s r^{n-1} b(r\sigma)\,dr$$

where $\sqrt{A}$ is a symmetric positive square root of A,

$$\leqq \int_{S^{n-1}} d\sigma \sqrt{\int_1^\rho <A\nabla e_0^\rho, \nabla e_0^\rho>(s\sigma) s^{n-1} ds} \sqrt{\int_1^\rho <A^{-1}\mathfrak{n}, \mathfrak{n}>(s\sigma) s^{1-n} [\int_1^s r^{n-1} b(r\sigma)\,dr]^2 ds}$$

$$\leqq \sqrt{\int_{1<|x|<\rho} <A\nabla e_0^\rho, \nabla e_0^\rho> dx} \sqrt{\int_{S^{n-1}} d\sigma \int_1^\rho <A^{-1}\mathfrak{n}, \mathfrak{n}>(s\sigma) s^{1-n} [\int_1^s r^{n-1} b(r\sigma)\,dr]^2 ds}.$$

$$\leqq \sqrt{\mathcal{E}(e_0^\rho, e_0^\rho)} \sqrt{\int_{S^{n-1}} d\sigma \int_1^\rho <A^{-1}\mathfrak{n}, \mathfrak{n}>(s\sigma) s^{1-n} [\int_1^s r^{n-1} b(r\sigma)\,dr]^2 ds}.$$

A simple application of the Dirichlet principle shows

$$\lim_{\rho\uparrow+\infty} \mathcal{E}_0(e_0^\rho, e_0^\rho) = \mathcal{E}_0(e_0, e_0) = C_0(B_n).$$

Thus we obtain

$$\int_{|x|>1} e_0(x) b(x)\,dx \leqq \sqrt{C_0(B_n)} \sqrt{\int_{S^{n-1}} d\sigma \int_1^{+\infty} <A(s\sigma)^{-1}\sigma, \sigma> s^{1-n} [\int_1^s r^{n-1} b(r\sigma)\,dr]^2 ds},$$

which is finite by the hypothesis. q.e.d.

§4. Some examples.

In order to illustrate Theorems 3 and 4, some examples are examined in this section.

Example 1. Consider the case $b = 1$ on R^n. If $a_{ij}(x) = O(|x|^2)$, then X_t is conservative. Indeed, it holds under the conditons that with positive constants C_1 and C_2, $A_+(r) \leqq C_1 r^2$ and $B_+(r) \leqq C_2 r^2$. Consequently, we have

$$\frac{\exp\{-\int_1^r \frac{2\sqrt{\alpha}}{\sqrt{B_+(s)}} ds\}}{\int_r^{+} s^{1-n} A_+(s)^{-1} ds} \leqq \frac{\exp\{-\int_1^r \frac{2\sqrt{\alpha}}{\sqrt{C_2}} \frac{ds}{s}\}}{\int_r^{+\infty} s^{1-n} \frac{ds}{s^2}}$$

$$= \text{const.}\ r^{n-2\sqrt{\frac{\alpha}{C_2}}} \to 0, \quad \text{as } r \to +\infty$$

for any positive α with $n < 2\sqrt{\frac{\alpha}{C_2}}$ Thus X_t is conservative.

Example 2. Let us assume that X_t is associated with the operator

$$L = \frac{1}{a(x)} \sum_{i=1}^{n} \frac{\partial}{\partial x_i} (a(x)\frac{\partial}{\partial x_i}).$$

Let $\bar{a}(r) = \int_{S^{n-1}} a(r\sigma)d\sigma$. If $\bar{a}(r)$ satisfies

$$(*) \qquad \lim_{r\to+\infty} \frac{\log[\int_r^{+\infty} \frac{ds}{s^{n-1}\bar{a}(s)}]}{r} > -\infty,$$

then X_t is conservative. In particular, if $\bar{a}(r) \leqq C_1 e^{C_2 r}$ holds with positive constants C_1 and C_2, then the above condition (*) is satisfied. Since $A_+(r) = \bar{a}(r)$ and $B_+(r) = 1$, it is easy to verify the condition in Theorem 3.

Example 3. Let $L = \sum_{i=1}^{n} \frac{\partial}{\partial x_i}(a_i(x)\frac{\partial}{\partial x_i})$. Suppose that with some positive constants C and $\alpha > 1$,

$$a_i(x) \geqq C|x|^{n+2}(\log|x|)^{\alpha}$$

for all sufficiently large $x \in R^n$. Then explosion of X_t is sure.

Example 4. Let $n \geqq 3$ and $a_{ij}(x) = \delta_{ij} = \begin{cases} 1, & \text{if } i=j \\ 0, & \text{otherwise.} \end{cases}$ If the following condition

$$\int_{|x|>1} \frac{b(x)dx}{|x|^{n-2}} < +\infty$$

is satisfied, then explosion of X_t is sure. Indeed, the hitting probability e_0 of X_t is given by $e_0(x) = \frac{1}{|x|^{n-2}}$ for $|x| \geqq 1$. Thus Corollary 1 gives the desired result.

References

1. Feller, W.: The parabolic differential equations and the associated semigroups of transformations. Ann. Math., 55, 468-519 (1952).
2. Fukushima, M.: Dirichlet forms and Markov processes. Kodansha, Tokyo (1980).
3. Hasminskii, R.Z.: Ergodic properties of recurrent diffusion processes and stabilization of the problem to the Cauchy problem for parabolic equations. T.V., 5, 196-214 (1960).
4. Ichihara, K.: Some global properties of symmetric diffusion processes. Publ. RIMS, Kyoto Univ., 14, 441-486 (1978).
5. Ichihara, K.: Explosion problems for symmetric diffusion processes. submitted to Trans. Amer. Math. Soc.
6. McKean, H.P.: Stochastic integrals. Academic Press, N.Y. (1969).
7. Mizohata, S.: The theory of partial differential equations. Cambridge Univ. Press, London (1973).
8. Tomisaki, M.: Dirichlet forms associated with direct product diffusion processes. Lecture notes in math., 923, Springer (1981).

EXTREMAL PROCESS AS A SUBSTITUTION FOR "ONE-SIDED STABLE PROCESS WITH INDEX 0"

Yuji Kasahara
Institute of Mathematics
University of Tsukuba
Sakura-mura
Ibaraki, 305 JAPAN

1. Introduction.

Let $X_\alpha = (X_\alpha(t))_{t \geq 0}$ $(X_\alpha(0) = 0)$ be a process with stationary independent increments with the Lévy measure $n_\alpha(dx) = I(x>0)\alpha x^{-\alpha-1}\,dx$ $(0 < \alpha < 1)$;

$$E[\, e^{-s X_\alpha(t)} \,] = e^{-t\Gamma(1-\alpha)s^\alpha}$$

$$= \exp t \int_0^\infty (e^{-sx} - 1)n_\alpha(dx), \quad s > 0,\ t \geq 0. \qquad (1.1)$$

X_α is called a one-sided stable process (or subordinator) with index α. As is well known one-sided stable process X_α plays an important role in various limit theorems and the aim of this note is to consider the extreme case as $\alpha \downarrow 0$ of those limit theorems. This extreme case is of interest because processes such as Cauchy process and 2-dimensional Brownian motion are sometimes involved. However, since "one-sided stable process with index 0" does not make sense (at least among non-trivial finite processes), we need to find a process that plays the role of X_α when $\alpha = 0$. In section 2 we will see that $e = (e(t))_t$ called an extremal process is a natural substitution for X_0 (in fact for $(X_\alpha)^\alpha$, $\alpha = 0$) when one is interested in Markov property or in self-similarity. (The process e no longer has independent increments but it is self-similar with index $H = 1$.) In sections 3 and 4 we will give examples to show how e plays the role of X_α when $\alpha = 0$: We will see that some of the well-known limit theorems which involves X_α $(0 < \alpha < 1)$ is still valid even in the extreme case $\alpha = 0$ if we modify the normalization and replace X_α by e.

2. Kotani's Lemma.

In this section we will study the asymptotic behavior of X_α as

$\alpha \downarrow 0$. For this purpose it is convenient to realize all X_α's $(0<\alpha<1)$ on a common probability space. Let $N(dt\,dx) = N(dt\,dx,\omega)$ be a Poisson random measure on $\{(t,x): t>0,\ x\in\mathbb{R}\setminus\{0\}\}$ with mean measure $dt\ x^{-2}dx$ and define for $0<\alpha<1$,

$$\tilde{X}_\alpha(t) = \int_0^{t+}\int_{x>0} x^{1/\alpha}\, N(du\,dx), \quad t \geq 0. \tag{2.1}$$

Notice that (2.1) may be rewritten as follows if we change the variable $(y = x^{1/\alpha})$.

$$\tilde{X}_\alpha(t) = \int_0^{t+}\int_{y>0} y\, N_\alpha(du\,dx), \quad t \geq 0, \tag{2.2}$$

where $N_\alpha(du\,dx)$ is the Poisson random measure with the mean measure $du\times y^{-\alpha-1}dy$ $(= du\ n_\alpha(dy))$. This implies that $\tilde{X}_\alpha$ is a process with stationay independent increments with the Lévy measure $n_\alpha(dy)$. By the definition of X_α this implies that $\tilde{X}_\alpha$ and X_α are equivalent in distribution. Therefore in what follows we may and do assume that (2.1) is the definition of X_α;

$$X_\alpha(t) = \int_0^{t+}\int_{x>0} x^{1/\alpha}\, N(du\,dx), \quad t \geq 0,\ (0<\alpha<1). \tag{2.3}$$

This may also be restated as

$$X_\alpha(t) = \sum_{0<s\leq t} (\Delta^+\xi(s))^{1/\alpha}, \qquad t \geq 0,\ (0<\alpha<1), \tag{2.4}$$

where ξ is the Cauchy process defined by

$$\xi(t) = \underset{\varepsilon\downarrow 0}{\text{l.i.p}} \int_0^{t+}\int_{|x|>\varepsilon} x\, N(du\,dx), \quad t \geq 0, \tag{2.5}$$

and

$$\Delta^+\xi(s) = \max\{\xi(s)-\xi(s-),\ 0\}, \quad s > 0.$$

The equivalence of (2.3) and (2.4) is clear because $N(du\,dx)$ is the same as the Poisson random measure defined by the discontinuities of ξ. Let us now define

$$e(t) = \sup_{0<s\leq t} \Delta^+\xi(s), \quad t>0 \quad (e(0)=0). \tag{2.6}$$

The finite-dimensional distribution of $e = (e(t))_t$ is given as follows; for $0 \leq t_1 < t_2 < \cdots < t_n$ and $0 \leq x_1 < x_2 < \cdots < x_n$, $n = 1, 2, \cdots$,

$$P[\, e(t_1) \leqq x_1 \,, \cdots , e(t_n) \leqq x_n \,]$$
$$= F(x_1)^{t_1} F(x_2)^{t_2 - t_1} \cdots F(x_n)^{t_n - t_{n-1}} \,, \qquad (2.7)$$

where $F(x) = e^{-1/x}$, $x > 0$.
To see this one needs only to notice that the left-hand side of (2.7) may be written as $P[\, N(\Gamma) = 0 \,]$ where $N(dt\,dx)$ is the same as before and $\Gamma = \cup_{k=1}^{n} [t_{k-1}, t_k) \times [x_k, \infty)$, $(t_0 = 0)$. Since $N(\Gamma)$ is a Poisson random variable this is equal to $\exp -E[N(\Gamma)] = \exp -\iint_\Gamma dt\, x^{-2} dx$. Thus we have (2.7). In general a process with the property (2.7) is called an <u>extremal process</u> with given distribution function $F(x)$ and appears in the extreme-value theory (see Dwass [2]). In the definition of extremal processes the distribution function $F(x)$ is arbitrary, but there is essentially only one extremal process because general extremal processes may be obtained from $e(t)$ via change of the scale.

Another expression for e is sometimes useful: Let $B = (B(t))_t$ be a one-dimensional standard Brownian motion starting at 0 and let

$$M(t) = \max_{0 \leqq s \leqq t} B(s) \quad , \quad t \geqq 0,$$

$$\ell(t) = \lim_{\varepsilon \downarrow 0} \frac{1}{2\varepsilon} \int_0^t I_{[0,\varepsilon)}(B(s))\, ds \,, \qquad t \geqq 0.$$

Then we have

$$e(t) \overset{d}{=} M(\ell^{-1}(t)) \qquad (2.8)$$

where $\ell^{-1}(t)$ denotes the right-continuous inverse of $\ell(t)$ and $\overset{d}{=}$ denotes the equivalenve in law of the processes. See [9] for the proof of (2.8). It is known that e is a Markov process (see [9] for the semigroup) and its inverse $e^{-1}(t)$ has independent increments (see [8]). It should be remarked that $e^{-1}(t)$ is exponentially distributed with expectation t for each fixed $t > 0$. (This is clear from (2.7).)

Let us now return to our problem. The question was to find a substitution for X_α when $\alpha = 0$, and our answer is that e is the desired one. The reason may be clear if we compare (2.4) and (2.6). Precisely speaking, e is a substitution not for X_α itself but for $(X_\alpha)^\alpha$. This difference is not essential as long as we are interested in the Markov property or in the self-similarity of X_α.

The following lemma is due to S. Kotani (private communication).

Lemma 1.

$$X_\alpha(t)^\alpha \xrightarrow{\mathcal{D}} e(t) \quad \text{in} \quad D[0, \infty) \quad \text{as} \quad \alpha \downarrow 0.$$

Here, $D[0, \infty)$ denotes the Skorohod function space endowed with the J_1-topology (see [7]) and $\xrightarrow{\mathcal{D}}$ denotes the convergence in distribution. S. Kotani's proof (see [9]) is interesting itself but in our formulation the assertion is almost obvious if we look at the definitions (2.4) and (2.6). Indeed, the assertion may be restated as

$$\Big(\sum_{s \leq t} (\Delta^+\xi(s))^{1/\alpha}\Big)^\alpha \xrightarrow{\mathcal{D}} \sup_{s \leq t} \Delta^+\xi(s) \quad \text{as} \quad \alpha \downarrow 0. \tag{2.9}$$

Therefore the proof can be carried out in the same manner as that of the fact that L^p-norm converges to the sup-norm as $p \to \infty$. The only problem is that $N(du\,dx)$ is infinite a.s. However, since $\int_0^T \int_{x>0} x^2\, N(dudx)$ is finite a.s. (1/2-stable law), we have that

$$\begin{aligned} e(t) &\leq X_\alpha(t)^\alpha \\ &\leq e(t)^{1-2\alpha}\Big(\int_0^{T+} \int_{x>0} x^2\, N(du\,dx)\Big)^\alpha, \quad 0 \leq t \leq T. \end{aligned}$$

Now it is easy to see that $X_\alpha(t)^\alpha$ converges to $e(t)$ uniformly for t $(0 \leq t \leq T)$ a.s.

Remarks. (i) A similar idea as (2.9) may be found in [12].
(ii) For the 1-dimensional marginal distribution we have the convergence of the density. The result is due to N. Cressie [11].

3. A limit theorem for sums of i.i.d. random variables.

Let $\xi_1, \xi_2, \cdots$ be nonnegative, independent random variables with common distribution function $F(x)$ belonging to the domain of attraction of stable law with index $\alpha (0 < \alpha < 1)$;

$$1 - F(x) \sim \frac{1}{\phi(x)}, \quad \text{as} \quad x \to \infty, \tag{3.1}$$

where $\phi(x)$ is a regularly varying function with index α;i.e., $\phi(x) = x^\alpha L(x)$ for some slowly varying $L(x)$. (Throughout $f(x) \sim g(x)$ means that $\lim f(x)/g(x) = 1$.)

Define $S_0 = 0$ and $S_n = \xi_1 + \xi_2 + \cdots + \xi_n$ $(n \geq 1)$. Then it is well known that for the asymptotic inverse $v(x)$ of $\phi(x)$ we have that

$$\frac{1}{v(n)} S_{[nt]} \xrightarrow{\mathcal{D}} X_\alpha(t) \quad \text{in} \quad D[0, \infty), \quad \text{as } n \to \infty . \tag{3.2}$$

Let us now consider the extreme case of (3.1)-(3.2) as $\alpha \downarrow 0$:

$$1 - F(x) \sim \frac{1}{\phi(x)} \quad \text{as} \quad x \to \infty ,$$

where $\phi(x) = L(x)$ is a slowly varying function. (3.3)

It is well known that, under this condition, every linear normalization $a_nS_n + b_n$ leads to a degenerate limiting distribution, and therefore, we do not have a meaningful statement if we formally let $\alpha \downarrow 0$ in (3.2). So let us rewrite (3.2) as follows.

$$\frac{1}{n} \phi(S_{[nt]}) \xrightarrow{\mathcal{D}} X_\alpha(t)^\alpha \text{ in } D[0, \infty), \quad \text{as} \quad n \to \infty. \tag{3.4}$$

To see the equivalence of (3.2) and (3.4) one need only to notice that $\phi(b_n)/\phi(a_n) \longrightarrow c^\alpha$, $n \to \infty$ if and only if $b_n/a_n \longrightarrow c$ provided that $a_n \longrightarrow \infty$, $n \to \infty$ and $c > 0$. Since $X_\alpha(t)^\alpha$ converges to $e(t)$ as $\alpha \downarrow 0$, we can naturally conjecture that (3.4) hold even if $\alpha = 0$ if we replace $X_\alpha(t)^\alpha$ by $e(t)$. In fact we have

Theorem 1. Let $L(x) \geqq 0$, $x \geqq 0$ be a nondecreasing function varying slowly at infinity. Then under the assumption (3.3), we have

$$\frac{1}{n} L(S_{[nt]}) \xrightarrow{\mathcal{D}} e(t) \quad \text{in} \quad D[0, \infty), \quad \text{as} \quad n \to \infty,$$

where $e = (e(t))_t$ is the same as before.

Remarks. (i) The convergence of 1-dimensional marginal distribution is due to D. A. Darling [10].
(ii) The convergence of the semi-group is proved by S. Watanabe [9].
(iii) Let ξ_k $(k = 1, 2, \cdots)$ be the length of the kth excursion of the simplest random walk on $\mathbb{Z}^2$ starting at (0,0). Then (3.3) is satisfied with $\phi(x) = (1/\pi) \log(1 + x)$.

Since the proof of Theorem 1 will be published elsewhere, we will only explain the story of the poof and we refer to [5] for details.

Let $\eta_1, \eta_2, \cdots$ be independent, identically distributed random variables with the distribution function $G(x) = 1 - \frac{1}{x}$ $(x \geqq 1)$ and $= 0$ $(x < 1)$.

Then the random measure

$$N_n(dt\,dx) = \sum_{k=1}^{\infty} \delta_{(k/n,\eta_k/n)}(dt\,dx)$$

converges in law to $I(x > 0)\ N(dt\,dx)$ where $N(dt\,dx)$ is the Poisson random measure given in section 1. Put $f(x) = F^{-1}(G(x))$. Then it is easy to see that

$$\frac{1}{n} L(S_{[nt]}) \stackrel{d}{=} \frac{1}{n} L\left(\int_0^{t+} \int_{x>0} f(nx)\ N_n(ds\,dx)\right). \tag{3.5}$$

Since it can also be proved that $f(nx)$ is close to $L^{-1}(nx)$ in a certain sense, we see that the left-hand side of (3.5) may be approximated by

$$\frac{1}{n} L\left(\int_0^{t+} \int_{x>0} L^{-1}(nx)\ N(ds\,dx) \right). \tag{3.6}$$

Like the typical case of $L(x) = \log(1+x)$, we easily see that (3.6) converges to $e(t) = \sup\{\Delta^+\xi(s): s \leq t\}$, which proves our theorem.

Remark. In the simplest case where $L(x) = \log(1 + x)$, we have a simpler proof: Observe that we clearly have

$$\frac{1}{n} L(M_{[nt]}) \leq \frac{1}{n} L(S_{[nt]}) \leq \frac{1}{n} L(nt\ M_{[nt]}), \tag{3.7}$$

where $M_0 = 0$ and $M_n = \max\{\xi_1, \xi_2, \cdots, \xi_n\}$, $n \geq 1$.
Since

$$\left|\frac{1}{n}L(nx) - \frac{1}{n}L(x)\right| \leq \frac{1}{n} \log n \longrightarrow 0 \quad \text{as} \quad n \to \infty,$$

we see that the three processes in (3.7) have the same limiting processes. Thus our problem may be reduced to the well-known result for the maxima of i.i.d. random variables $L(\xi_1), L(\xi_2), \cdots$. (see e.g. [13]). However, in general the first two processes in (3.7) converge to $e = (e(t))$ but the last one does not necessarily have the same limit process. $L(x) = \exp\sqrt{\log(1+x)}$ is a counter example. In this case the limit process of the last one is $\sqrt{e}\ e(t)$.

4. Occupation times of Markov processes.

Let $(X_t)_{t \geq 0}$ be a temporally homogeneous Markov process with values in a measurable space and let $f(x) \geq 0$ be a bounded measurable function on it. Then the random variable $\int_0^t f(X_s)\,ds$ is called the occupation time because if $f(x) = I_E(x)$ then this random variable becomes the time $(X_t)_{t \geq 0}$ spent in the set E. There have been many works on the asymptotic behavior of this random variable as $t \to \infty$ and among them the following result due to D. A. Darling and M. Kac [14] is well known.

Suppose that for some $C > 0$ and $0 \leq \alpha < 1$,

$$E_x[\int_0^\infty e^{-st} f(X_t)\,dt] \sim C\, s^{-\alpha} L(1/s) \quad \text{as} \quad s \downarrow 0 \tag{4.1}$$

holds uniformly for $x \in \{y: f(y) > 0\}$. Then the law of

$$\frac{1}{C\, t^\alpha L(t)} \int_0^t f(X_s)\,ds$$

converges weakly to the Mittag-Leffler distribution with index α as $t \to \infty$.

Notice that for every fixed x, (4.1) is equivalent to

$$E_x[\int_0^t f(X_s)\,ds] \sim \frac{C}{\Gamma(\alpha+1)} t^\alpha L(t) \quad \text{as} \quad t \to \infty, \tag{4.2}$$

or equivalently, to

$$E_x[\frac{1}{t^\alpha L(t)} \int_0^t f(X_s)\,ds] \sim \frac{C}{\Gamma(\alpha+1)}, \quad \text{as} \quad t \to \infty. \tag{4.3}$$

A functional limit theorem was proved by N. H. Bingham [1] : If $0 < \alpha < 1$, then under the Darling-Kac condition it holds that

$$\frac{1}{\lambda^\alpha L(\lambda)} \int_0^{\lambda t} f(X_s)\,ds \xrightarrow{\mathcal{D}} \text{const.}\, X_\alpha^{-1}(t) \quad \text{as } \lambda \to \infty \tag{4.4}$$

Here, of course $X_\alpha^{-1}(t)$ denotes the right-continuous inverse process of the one-sided stable $X_\alpha(t)$.

Notice that, in the Darling-Kac theorem, the extreme case $\alpha = 0$ is included and the limiting distribution is the exponential distribution, while in Bingham's functional limit theorem we have to exclude this case. This may sound a little curious but the following observation will explain the situation. Suppose we have a limit theorem like (4.4) with $\alpha = 0$ for some limiting process $Z = (Z(t))_t$, then it is a standard argument to conclude that Z is self-similar with index $H = 0$; i.e, $Z(ct) \stackrel{d}{=} Z(t)$

for every $c > 0$. Since Z is monotone, this implies that Z is degenerate in the sense that Z does not depend on t; i.e, $Z(t) \equiv \zeta$, $t > 0$, where ζ is an exponential random variable (by Darling-Kac's theorem).

Thus if $\alpha = 0$ then we have a meaningful limit theorem for one-dimensional distributions but the functional limit theorem is less interesting. However, the case $\alpha = 0$ is still of interest because as we will expain later this case includes the Cauchy process and the two-dimensional standard Brownian motion. (For one-dimensional marginal distribution the result is well known as Kallianpur-Robbins' theorem.) Therefore, let us try to find a nontrivial limit theorem for the case $\alpha = 0$.

Now let us rewrite (4.4) as follows;

$$\frac{1}{\lambda} \int_0^{\phi^{-1}(\lambda t)} f(X_s)\, ds \xrightarrow{\mathcal{D}} \text{const.} \cdot X_\alpha^{-1}(t^{1/\alpha}), \quad \lambda \to \infty. \quad (4.5)$$

where $\phi(t) = t^\alpha L(t)$.
This is immediate from (4.4) because $\phi^{-1}(t)$ varies regularly with index $1/\alpha$ and hence $\phi^{-1}(\lambda t) \sim \phi^{-1}(\lambda) t^{1/\alpha}$ as $\lambda \to \infty$. Observe that the limit process $X_\alpha^{-1}(t^{1/\alpha})$ in (4.5) is the inverse of $X_\alpha(t)^\alpha$ which appeared in the previous sections. Since $X_\alpha(t)^\alpha \longrightarrow e(t)$ as $\alpha \downarrow 0$, letting $\alpha \downarrow 0$ in (4.5) we formally have that

$$\frac{1}{\lambda} \int_0^{\phi^{-1}(\lambda t)} f(X_s)\, ds \longrightarrow \text{const.} \cdot e^{-1}(t), \quad \lambda \to \infty \quad (4.6)$$

provided (4.1) (or, (4.2)) holds with $\alpha = 0$ and $\phi(t) = L(t)$.

In fact we can prove (4.6) in the sense of all finite-dimensional marginal distributions. (The convergence in distribution over the function space $D[0, \infty)$ does not hold --- the left-hand side of (4.6) is a continuous process if $\phi^{-1}(t)$ is continuous but the limit $e(t)$ is discontinuous, which proves that the convergence in J_1 never occurs, since $C[0, \infty)$ is closed under J_1-topology.)

<u>Theorem 2.</u> Let $L(t) \geq 0$ $(t \geq 0)$ be a monotone increasing function tending to infinity and varying slowly as $t \to \infty$. If (4.2) holds uniformly for $x \in \{y: f(y) > 0\}$ with $\alpha = 0$ then

$$\frac{1}{\lambda} \int_0^{L^{-1}(\lambda t)} f(X_s)\, ds \xrightarrow[\text{f.d.}]{} C\, e^{-1}(t) \quad \text{as } \lambda \to \infty,$$

where $\xrightarrow[\text{f.d.}]{}$ denotes the weak convergence of all finite-dimensional marginal distributions.

For the proof of Theorem 2, the moment method of Darling-Kac [14] and Bingham [1] is still applicable: We prove the convergence of the Laplace transform of every moment of $(1/\lambda)\int_0^{L^{-1}(\lambda t)} f(X_s)ds$. The only one difficulty arises from the time-inhomogeneity of this process; we need an elaborate argument and tedious computation to evaluate the limit of the moments. See [3] for the complete proof.

Example 4.1. ([6]) (2-dimensional Brownian motion.)

Let $(X_t)_{t \geqq 0}$ be a 2-dimensional standard Brownian motion and let $f(x)$, $x \in \mathbb{R}^2$ be a bounded measurable function vanishing outside a compact set. Then,

$$E_x[\ f(X_s)\ ] = \int_{\mathbb{R}^2} \frac{1}{2\pi t} e^{-|x-y|^2/2t} f(y)dy$$

$$\sim \quad C/t \qquad \text{as} \quad t \to \infty ,$$

where $C = \int_{\mathbb{R}^2} f(y)\ dy/2\pi$.

Therefore, we have that

$$E_x[\ \int_0^t f(X_s)ds\ ] \quad \sim \quad C \log t \quad \text{as} \quad t \to \infty .$$

Thus we have from Theorem 2 that, as $\lambda \to \infty$,

$$\frac{1}{\lambda} \int_0^{e^{\lambda t}} f(X_s)\ ds \xrightarrow[\text{f.d.}]{} \frac{1}{2\pi} \int f(y)dy\ e^{-1}(t).$$

(In Theorem 2 we assumed that $f(x) \geqq 0$ which condition may be removed in this example.)

Remark. Theorem 2 is still valid for discrete-time Markov processes (the proof needs an obvious modification). Therefore, for the simplest random walk on the plane we have a result similar to Example 4.1. (In this case $C = (1/\pi) \sum f(n,m)$.)

Example 4.2. (Cauchy process).

Let $(X_t)_{t \geqq 0}$ be the (symmetric) Cauchy process and let $f(x)$, $x \in \mathbb{R}$ be a bounded measurable function with a compact support. Then, since

$$E_x[\ f(X_t)\ ] = \frac{t}{\pi} \int_{\mathbb{R}} f(y)/\{(y-x)^2 + t^2\}dy$$

$$\sim \frac{1}{\pi} \int_{\mathbb{R}} f(y)\ dy\ \frac{1}{t} \qquad \text{as} \quad t \to \infty \quad ,$$

we see that

$$E_x[\ \int_0^t f(X_s)\ ds] \ \sim \ \frac{1}{\pi} \int_{\mathbb{R}} f(y)\ dy\ \log t \quad \text{as} \quad t \to \infty.$$

Therefore, we have that

$$\frac{1}{\lambda} \int_0^{e^{\lambda t}} f(X_s)\ ds \ \xrightarrow[\ ]{} {}_{f.d.} \ \frac{1}{\pi} \int_{\mathbb{R}} f(y)\ dy\ \ e^{-1}(t) \quad \text{as } \lambda \to \infty.$$

(Here again we may drop the condition that $f(x) \geqq 0$.)

Remark. In the above two examples we dropped the assumption that $f(x)$ is nonnegative (without explanation). In general it is sometimes important to consider the case where $f(x)$ may take negative values because when we want to compare the occupation times of two different sets, we need to consider the case where $f(x) = I_E(x) - I_F(x)$. However, in many cases we may treat this case in a similar way (with additional assumptions) as long as (4.1) holds with $C > 0$. But if $C = 0$ then the limiting process degenerates (see Examples 4.1 and 2) and therefore we need to change the normalization in order to have a non-trivial limit processes. See [4] for this problem.

REFERENCES

[1] N.H.Bingham: Limit theorems for occupation times of Markov processes. Z. Wahrsh. verw Geb. 17(1971), 1-22.

[2] M.Dwass: Extremal processes. Ann. Math. Stat. 35(1964), 1718-1725.

[3] Y.Kasahara: A limit theorem for slowly increasing occupation times. Ann. Probab. 10(1982), 728-736.

[4] Y.Kasahara: Another limit theorem for slowly increasing occupation times. J. Math. Kyoto Univ. 24(1984), 507-520.

[5] Y.Kasahara: A limit theorem for sums of i.i.d. random variables with slowly varying tail probability. (to appear)

[6] Y.Kasahara and S.Kotani: On limit processes for a class of additive functionals of diffusion processes. Z. Wahrsh. verw Geb. 49(1979), 133-153.

[7] T.Lindvall: Weak convergence of probability measures and random functions in the function space D[0, ∞), J.Appl. Probab. 10(1973), 109-121.

[8] S.I.Resnick: Inverses of extremal processes. Adv. Appl. Probab. 6 (1974), 392-406.

[9] S.Watanabe: A limit theorem for sums of non-negative i.i.d. random variables with slowly varying tail probabilities. Proc. 5th International Conf. Multivariete Anal. (1980), 249-261. North-Holland, Amsterdam.

[10] D.A.Darling: The influence of the maximum term in the addition of independent random variables. Trans. Amer. Math. Soc. 73(1952), 95-107.

[11] N.Cressie: A note on the behavior of the stable distribution for small index α. Z. Wahrsh. verw Geb. 33(1975), 61-64.

[12] W. Vervaat: On a stochastic difference equation and a represntation of non-negative infinitely divisible random variables. Adv. Appl. Probab. 11(1979), 750-783.

[13] J.Lamperti: On extreme order statistics. Ann. Math. Statist. 35 (1964), 1726-1737.

DIFFUSION MODEL OF POPULATION GENETICS INCORPORATING GROUP SELECTION, WITH SPECIAL REFERENCE TO AN ALTRUISTIC TRAIT

Motoo Kimura

National Institute of Genetics
Mishima, 411 Japan

Abstract

In order to investigate under what conditions an altruistic trait evolves through group selection, the following diffusion model is formulated. Consider a species consisting of an infinite number of competing groups (demes) each having a constant number of reproducing members and in which mating is at random. Then consider a gene locus and assume a pair of alleles $\underline{A}$ and $\underline{A}'$, where $\underline{A}'$ is the "altruistic allele." Let x be the relative frequency of $\underline{A}'$ within a deme, and let $\phi = \phi(x; t)$ be the density function of x at time t such that $\phi(x; t)\Delta x$ represents the fraction of demes whose frequency of $\underline{A}'$ lies in the range $(x, x + \Delta x)$. Then, we have

$$\frac{\partial\phi}{\partial t} = \frac{1}{2}\frac{\partial^2}{\partial x^2}\{V_{\delta x}\phi\} - \frac{\partial}{\partial x}\{M_{\delta x}\phi\} + c(x - \bar{x})\phi,$$

where $M_{\delta x}$ and $V_{\delta x}$ stand for the mean and variance of the change in x per generation (due to mutation, migration, individual selection and random sampling of gametes) within demes, and c is a positive constant (called the coefficient of interdeme competition) and $\bar{x}$ is the mean of x over the species, i.e. $\bar{x} = \int_0^1 x\phi dx$. By studying the above diffusion equation at steady state ($\partial\phi/\partial t = 0$), a condition is obtained for group selection to prevail over individual selection in the evolution of an altruistic trait.

Introduction

I would like to thank Professor Hida for inviting me to talk in this "Conference on Stochastic Processes and their Applications." As a theoretical population geneticist, I have worked, over the last 30 years, on stochastic processes of gene frequencies in finite populations in terms of diffusion models, that is, by using the diffusion equation method. Since I started my career as a botanist, my knowledge, as well as ability, in mathematics is quite limited. So, I am somewhat embarrassed by finding myself in a situation where I have to talk on mathematical models in front of an audience full of great experts in mathematical probability.

In the past, population genetics has offered many interesting problems for mathematicians to work on, so, I hope that my present talk will stimulate some of you to investigate further my model of group selection in more rigorous mathematical terms.

In the field of population genetics, diffusion models have been used quite successfully in treating stochastic processes of gene frequencies in finite populations (Kimura, 1964; Crow and Kimura, 1970; Kimura and Ohta, 1971). Particularly, diffusion models have been of great value for developing theoretical population genetics at the molecular level (Kimura, 1971; Nei, 1975; for review, see Kimura, 1983a). In all these treatments, it is customary to assume that natural selection acts through survival and reproduction of individuals. In other words, we assume individual selection as envisaged by Darwin (1859).

For the Darwinian notion of evolution by natural selection, however, an altruistic trait presents an apparent difficulty, because the practice of altruism for the benefit of others must decrease the fitness of the altruistic individual. The problem of how altruistic traits evolved seems to have attracted much attention, particularly since the publication of E. O. Wilson's (1975) book on "sociobiology."

One plausible explanation for the evolution of such a trait is to resort to group selection; namely, if a species consists of a large number of more or less isolated groups (or "demes") competing with each other, a mutant gene that induces an individual to sacrifice itself for the good of its group, under suitable conditions, will spread to the species.

This is because groups containing such an altruistic allele in higher frequency will tend to win in competition with other groups and proliferate, thereby helping such an allele to increase its frequency in the species, even if the altruistic allele is selected against within each deme.

In what follows, I shall present a diffusion model for treating intergroup selection in addition to conventional individual selection, mutation, migration and random drift. I shall then propose an index (to be denoted by D_K) which serves as a good indicator for predicting which of the two, that is, group selection or individual selection, prevails, when they counteract each other.

Diffusion equation incorporating group selection

Let us assume a hypothetical population (species) consisting of an infinite number of competing subgroups (demes) each with N breeding diploid individuals (more precisely, with the effective size N; see p. 109 of Crow & Kimura, 1970, for the meaning of the effective size). Consider a particular gene locus and assume a pair of alleles A and A'. We shall refer to A' as the "altruistic allele." This is an allele which lowers the Darwinian fitness of individuals containing it, but is helpful for demes to win in competition and proliferate.

Let us denote by x the relative frequency (proportion) of the altruistic allele A' within a deme, and consider the frequency distribution of x among the demes making up the species. Let $\phi(x; t)$ be the distribution function of x at time t such that $\phi(x; t)\Delta x$ represents the fraction of demes whose frequency of A' lies in the range $(x, x + \Delta x)$, where $0 < x < 1$.

We assume that random mating takes place within each deme, and mutation and individual selection cause gene frequency change as explained below.

Mutation occurs from A to A' at the rate v' per generation, and, in the reverse direction at the rate v. Thus, v' is the rate at which the altruistic allele (A') is produced, and v is the back mutation rate. Therefore, the rate of change in x per generation by mutation is $v'(1 - x) - vx$. We denote by s' the selection coefficient against A', and assume "genic selection" (i.e. semidominance or "no dominance" in fitness), so that the relative fitnesses of the three genotypes A'A', A'A and AA are respectively $1 - 2s'\Delta t$, $1 - s'\Delta t$ and 1, for a short time interval of length Δt. Then, under random mating within demes, the change of x during this time is

$$\frac{x - [2x^2 + x(1 - x)]s'\Delta t}{1 - 2xs'\Delta t} - x,$$

which reduces to $-s'x(1 - x)\Delta t$ at the limit $\Delta t \to 0$. Thus, taking one generation as the unit length of time, the rate of change in x due to individual selection is $-s'x(1 - x)$.

Migration is assumed to occur following Wright's (1931) island model, namely, we assume that, in each generation, individual demes contribute emigrants to the entire gene pool of the species at the rate m, and then receive immigrants from that pool at the same rate. Then, if $\bar{x}$ is the mean gene frequency of $\underline{A}'$ in the entire species, the change per generation of x in a given deme by migration is $m(\bar{x} - x)$.

Let us now consider the effect of interdeme selection. We shall denote by c(x) the coefficient of interdeme selection. This represents the rate at which the number of demes belonging to the gene frequency class x change through interdeme competition. This means that during a short time interval of length Δt, the change of $\phi(x;\, t)$ is

$$\Delta\phi = \frac{1 + c(x)\Delta t}{1 + \overline{c(x)}\Delta t}\,\phi - \phi, \tag{1}$$

where ϕ stands for $\phi(x;\, t)$ and $\overline{c(x)}$ is the average of c(x) over the entire array of demes in the species, namely,

$$\overline{c(x)} = \int_0^1 c(x)\phi(x;\, t)dx. \tag{1a}$$

At the limit $\Delta t \to 0$, we have

$$\Delta\phi = [c(x) - \overline{c(x)}]\phi\Delta t. \tag{2}$$

In the following, we shall treat the case in which c(x) is linear so that (2) becomes

$$\Delta\phi = c(x - \bar{x})\phi\Delta t, \tag{3}$$

where c is a positive constant.

Finally, we consider random change of x through sampling of gametes in reproduction. The change of x per generation through this cause has the mean zero and variance $x(1 - x)/(2N)$, which is the binomial variance when 2N gametes are sampled from a deme to form the next generation.

Combining all the above evolutionary factors, the diffusion equation for $\phi(x;\, t)$ is

(4) $$\frac{\partial\phi(x;\ t)}{\partial t} = \frac{1}{2}\frac{\partial^2}{\partial x^2}\left\{V_{\delta x}\phi(x;\ t)\right\} - \frac{\partial}{\partial x}\left\{M_{\delta x}\phi(x;\ t)\right\} + c(x - \bar{x})\phi(x;\ t),$$

where

(4a) $$V_{\delta x} = x(1 - x)/(2N)$$

is the variance of the change in x per generation due to random sampling of gametes,

(4b) $$M_{\delta x} = v'(1 - x) - vx + m(\bar{x} - x) - s'x(1 - x)$$

is the mean change per generation of x within demes due to the joint effect of mutation, migration and individual selection, c is the coefficient of interdeme competition, and $\bar{x}$ is the mean frequency of the altruistic allele over the species, i.e.,

(4c) $$\bar{x} = \int_0^1 x\phi(x;\ t)dx.$$

The diffusion equation (4) may also be expressed as

(5) $$\frac{\partial\phi}{\partial t} = L(\phi) + c(x - \bar{x})\phi,$$

where ϕ stands for $\phi(x;\ t)$ and

(5a) $$L = \frac{1}{2}\frac{\partial^2}{\partial x^2}V_{\delta x} - \frac{\partial}{\partial x}M_{\delta x}$$

is the Kolmogorov forward differential operator commonly used in population genetics (See, Kimura, 1964; p. 372 of Crow and Kimura, 1970; p. 167 of Kimura and Ohta, 1971). Note that $L(\phi)$ represents the change in the distribution of gene frequency (ϕ) caused by gene frequency changes inside demes, while $c(x - \bar{x})\phi$ represents the change of ϕ through competition between demes (i.e., group selection).

For the purpose of deriving expressions for the moments of x, the following equation is useful. Let f(x) be a polynomial of x, or a suitable continuous function of x. Then, using the same method as used by Ohta and Kimura (1971) to derive their Equation (A3), we obtain

$$(6) \qquad \frac{\partial E\{f(x)\}}{\partial t} = E\left\{ \frac{V_{\delta x}}{2} \frac{\partial^2 f(x)}{\partial x^2} + M_{\delta x} \frac{\partial f(x)}{\partial x} + c(x - \bar{x})f(x) \right\},$$

where E stands for the operator of taking expectation with respect to the distribution ϕ, so that, for example,

$$(6a) \qquad E\{f(x)\} = \int_0^1 f(x)\phi(x, t)dt.$$

Equation (6) reduces to Equation (A3) of Ohta and Kimura (1971) when $c = 0$, i.e., if group selection is absent.

Distribution at steady state

I shall now investigate the steady state distribution which will be realized when a statistical equilibrium is reached under the joint effects of recurrent mutation, migration, individual selection, random genetic drift, and group selection. We shall denote such a steady state distribution by $\phi(x)$. For this state, we have

$$\frac{\partial \phi(x;\ t)}{\partial t} = 0$$

in Equation (4), and therefore, we get

$$(7) \qquad \frac{1}{2} \frac{d^2}{dx^2} \{V_{\delta x}\phi(x)\} - \frac{d}{dx}\{M_{\delta x}\phi(x)\} + c(x - \bar{x})\phi(x) = 0.$$

To solve this differential equation, let

$$(8) \qquad \phi(x) = \phi_0(x)\psi(x),$$

and choose $\phi_0(x)$ so that it represents the steady state distribution attained if group selection were absent, i.e., if $c = 0$. Note that $\phi_0(x)$ satisfies the zero probability-flux condition (see Kimura, 1964; p. 170 of Kimura and Ohta, 1971), i.e.,

$$(9) \qquad -\frac{1}{2}\frac{d}{dx}\{V_{\delta x}\phi_0(x)\} + M_{\delta x}\phi_0(x) = 0,$$

$(0 < x < 1)$. This leads to Wright's (1938) distribution,

$$(10) \qquad \phi_0(x) = \frac{C}{V_{\delta x}} \exp\left\{2\int \frac{M_{\delta x}}{V_{\delta x}}\, dx\right\},$$

where C is a constant determined so that

$$(10a) \qquad \int_0^1 \phi_0(x)dx = 1.$$

Substituting (4a) and (4b) in (10), we obtain

$$(11) \qquad \phi_0(x) = \text{Const. } e^{-S'x}\, x^{\alpha-1}\,(1 - x)^{\beta-1}$$

where $S' = 4Ns'$, and, $\alpha = V' + M\bar{x}$ and $\beta = V + M(1 - \bar{x})$ in which $V' = 4Nv'$, $V = 4Nv$ and $M = 4Nm$ $(0 < x < 1)$. Since the constant term (Const.) is simply a normalizing factor so that all the frequency classes add up to unity, I shall neglect this in the following treatment.

Applying condition (9) to Equation (7) in which $\phi(x) = \phi_0(x)\psi(x)$, we obtain the differential equation,

$$(12) \qquad \frac{V_{\delta x}}{2}\frac{d^2\psi(x)}{dx^2} + M_{\delta x}\frac{d\psi(x)}{dx} + c(x - \bar{x})\psi(x) = 0,$$

where

$$(12a) \qquad \bar{x} = \frac{\int_0^1 x\phi_0(x)\psi(x)dx}{\int_0^1 \phi_0(x)\psi(x)dx}.$$

Since $\psi(x)$ must be a constant for $c = 0$, we attempt to solve Equation (12) under the condition that $\psi(x)$ remains finite and continuous throughout the closed interval [0, 1], including the two singular points, $x = 0$ and $x = 1$. In particular, we let $\psi(0) = 1$.

Substituting (4a) and (4b) in (12), we have

$$x(1-x)\frac{d^2\psi}{dx^2} + [V'(1-x) - Vx + M(\bar{x}-x) - S'x(1-x)]\frac{d\psi}{dx} + C(x-\bar{x})\psi = 0, \tag{13}$$

where ψ stands for $\psi(x)$, and $V' = 4Nv'$, $V = 4Nv$, $M = 4Nm$, $S' = 4Ns'$ and $C = 4Nc$.

Let $\psi(x) = e^{y(x)}$, then (13) becomes

$$x(1-x)\left[\frac{d^2y}{dx^2} + \left(\frac{dy}{dx}\right)^2\right] + [V'(1-x) - Vx + M(\bar{x}-x) - S'x(1-x)]\frac{dy}{dx} + C(x-\bar{x}) = 0, \tag{14}$$

$(0 \leqq x \leqq 1)$.

Letting $x \to 0$ and noting that d^ny/dx^n remains finite at $x = 0$, $(n = 1, 2, \cdots)$, we obtain

$$y'(0) = \frac{C\bar{x}}{V' + M\bar{x}}, \tag{15a}$$

$$y''(0) = \frac{1}{1+V'+M\bar{x}}\{(V'+V+M+S')y'(0) - [y'(0)]^2 - C\}, \tag{15b}$$

etc., where primes on y denote differentiation with respect to x. In terms of these, $y(x)$ may be expressed, in the neighborhood of $x = 0$, by the Taylor series.

$$y(x) = y'(0)x + \frac{1}{2}y''(0)x^2 + \cdots . \tag{16}$$

Note that $y(0) = 0$, because we assume $\psi(0) = e^{y(0)} = 1$.

Similarly, letting $x \to 1$, we obtain

$$y'(1) = \frac{C(1-\bar{x})}{V + M(1-\bar{x})}, \tag{17a}$$

$$y''(1) = \frac{1}{1+V+M(1-\bar{x})}\left\{(-V'-V-M+S')y'(1)-[y'(1)]^2 + C\right\}, \tag{17b}$$

etc. Thus, in the neighborhood of $x = 1$, we have

$$(18) \qquad y(x) = y(1) - y'(1)(1 - x) + \frac{1}{2}y''(1)(1 - x)^2 + \cdots .$$

These Taylor series are useful in obtaining an approximate analytical solution to our problem, particularly when mutation terms in Equation (14) can be neglected. They are also useful to solve Equation (14) numerically, thereby obtaining an exact numerical value of $\bar{x}$ and the shape of the steady state distribution.

Then, the required solution of the original equation (Equation 7) can be expressed in terms of $y(x)$ as follows.

$$(19) \qquad \phi(x) = \text{Const.} \exp\{y(x) - S'x\} \cdot [x^{\alpha-1}(1 - x)^{\beta-1}].$$

In applying a numerical method to obtain $y(x)$ by solving Equation (14) we must be careful in the neighborhood of the two boundaries, $x = 0$ and $x = 1$, which are the singular points of this differential equation. Thus, in deriving the numerical results to be presented later, I used the following method. For a small value of x, say $x < 0.05$, Equation (16) was used to obtain values of $y(x)$. Beyond such a small value of x Runge-Kutta methods were employed to integrate Equation (14) step by step to a certain point near 1, say $x = 0.9$, beyond which Equation (18) was incorporated to make sure of the process of numerical integration. Since $\bar{x}$ is unknown, I gave a trial value for $\bar{x}$, and, after obtaining the numerical solution of $y(x)$, I computed the value of $\bar{x}$ using Equation (12a). This process was repreated until the trial and the computed values of $\bar{x}$ were sufficiently close to each other.

Useful approximations

In a realistic situation in which mutation rates v and v' are much smaller than the migration rate m, we may neglect the terms containing V' and V in Equation (14), so that it reduces to

$$(20) \qquad x(1 - x)\left[\frac{d^2y}{dx^2} + \left(\frac{dy}{dx}\right)^2 - S'\frac{dy}{dx}\right] + \left(M\frac{dy}{dx} - C\right)(\bar{x} - x) = 0.$$

In this case, the power series solution in the neighborhood of $x = 0$ is

(21a) $$y(x) = Rx + \frac{1}{2}\frac{R(S' - R)}{1 + M\bar{x}}x^2 + \cdots,$$

and that in the neighborhood of $x = 1$ is

(21b) $$y(x) = y(1) - R(1 - x) + \frac{1}{2}\frac{R(S' - R)}{1 + M(1 - \bar{x})}(1 - x)^2 + \cdots,$$

where

$$R = C/M = c/m.$$

Equation (21a) suggests that

(22) $$y(x) = Rx$$

is a good approximation to the solution of (20) when $|S' - R|$ is small. In fact, it can be shown that (22) is the exact solution of (20) when $S' - R = 0$.

From the biological standpoint, this case is of particular interest, because, if $S' = R$, we have $y(x) - S'x = 0$ from (22), and therefore the exponent in Equation (19) vanishes; namely, the disadvantage coming from individual selection is cancelled by the advantage based on group selection. In other words, the altruistic allele behaves as if selectively neutral, giving $\bar{x} = v'/(v + v')$.

This suggest (Kimura, 1983b) that the index

(23) $$D = R - S' = c/m - 4Ns'$$

serves as a good indicator for predicting which of the two modes of selection predominates; if $D > 0$, group selection overrides individual selection, but if $D < 0$ individual selection wins over group selection.

Around this neutral point (i.e., when $|D|$ is small), we have

(24) $$\phi(x) = e^{Dx}x^{\alpha-1}(1 - x)^{\beta-1}$$

as a good approximation to the gene frequency distribution.

Substituting this approximation to $\phi(x)$ for $\phi_0(x)\psi(x)$ in (12a), and incorporating further approximations, such as $e^{Dx} = 1 + Dx$, we obtain

(25) $$\bar{x} = \frac{\alpha}{\alpha + \beta}\left[1 + D\frac{\beta}{(\alpha + \beta)(1 + \alpha + \beta)}\right].$$

If we note in addition that the mutation rates are much lower than the migration rate, i.e., $v' + v << m$, we obtain

(26) $$\bar{x} = \frac{1}{2BD}\left[BD - 1 + \sqrt{(1 - BD)^2 + 4\hat{x}_0 BD}\right],$$

where $B = M/[(M + 1)(V + V')]$ and $\hat{x}_0 = V'/(V + V')$. Note that $\hat{x}_0$ is the equilibrium frequency of A' attained if the alleles were selectively neutral.

As will be shown in the next section, the transition between the two states, that is, the state in which individual selection predominates and that in which group selection predominates, occurs rather abruptly as D changes from negative values to positive values.

A more accurate indicator for deciding which of these two types of selection predominates may be obtained by incorporating mutation terms in the above treatment. We note that $y(x) - S'x$, which appears as the exponent of Equation (19), is approximately

$$\left\{\frac{C\hat{x}_0}{V' + M\hat{x}_0} - S'\right\}x$$

in the neighborhood of the neutral point $\hat{x}_0 = V'/(V + V')$ (see Equation 15a). This suggests that an index,

$$D_K = \frac{C}{V'/\hat{x}_0 + M} - S'$$

or

(27) $$D_K = \frac{c}{v + v' + m} - 4Ns'$$

serves as a more accurate indicator for our purpose. In fact, numerical analysis shows that this is so. Note that D_K reduces to D if we neglect $v + v'$ in the denominator of the right-hand side of Equation (27).

Numerical studies

We can compute the frequency distribution at steady state, i.e., $\phi(x)$ in the following way. For a given set of parameters, v, v', m, s', c and N, we first give a trial value for the mean $\bar{x}$, and compute $y(x)$ by numerically solving the differential equation (14). Then, combining $\psi(x) = e^{y(x)}$ and $\phi_0(x)$, i.e. equation (11), we compute a new value of $\bar{x}$ by using formula (12a). This process is repeated until the trial value of $\bar{x}$ and the computed value of $\bar{x}$ become sufficiently close.

The final trial value $\bar{x}$ thus obtained serves to derive the steady state distribution $\phi(x)$. The variance and the higher moments can be obtained from this distribution. They can also be obtained from equations (6) by putting $f(x) = x^n$, $(n = 1, 2, 3, \cdots)$, and assuming $\partial E\{f(x)\}/\partial t = 0$, if the correct value of $\bar{x}$ is known.

I am indebted to Dr. N. Shimakura (personal communication) for his suggestion that $\bar{x}$ at steady state may also be computed by using a continued fraction method as explained below. Let $f(x) = x^n$ in Eq. (6), then we have a set of equations for the moments:

$$\text{(28)} \quad \frac{d\mu_n'}{dt} = (ns' + c)\mu_{n+1}' - \left[\frac{n(n-1)}{4N} + n(v' + m\bar{x}) + c\bar{x}\right]\mu_n' - \left[\frac{n(n-1)}{4N} + (v' + v + m + s')n\right]\mu_{n-1}'$$

where

$$\mu_n' = \int_0^1 x^n \phi(x;\ t)dt$$

and $\mu_0' = 1$ $(n = 1, 2, \cdots)$.

At steady state in which $d\mu_n'/dt = 0$ for all n, we have

$$\text{(29)} \quad (nS' + C)\mu_{n+1}' - [C\bar{x} + n(V + V' + S' + M + n - 1)]\mu_n' + n(V' + M\bar{x} + n - 1)\mu_{n-1}' = 0.$$

Let $R_n = \mu_n'/\mu_{n-1}'$, then this leads to the continued fraction

$$\text{(30)} \quad \bar{x} = R_1 = \frac{a_1(\bar{x})}{b_1(\bar{x})} \frac{}{-} \frac{a_2(\bar{x})}{b_2(\bar{x})} \frac{}{-} \cdots ,$$

where

$$a_n(\bar{x}) = n(V' + M\bar{x} + n - 1)/(nS' + C)$$

and

$$b_n(\bar{x}) = [n(V + V' + S' + M + n - 1) + C\bar{x}]/(nS' + C).$$

For a given set of parameters, V, V', M, S' and C, and for a trial value of $\bar{x}$, we can compute a new value of $\bar{x}$ by Eq. (30), and this may be repeated until the trial value and the computed value of $\bar{x}$ agree with a desired accuracy. Note that in diffusion models, parameters giving mutation, migration and selection all enter as product with N, namely, as $V' = 4Nv'$, $V = 4Nv$, $M = 4Nm$, $S = 4Ns'$ and $C = 4Nc$.

In Table 1, an example is presented in which the intensity of individual selection is varied (S' or 4Ns' is changed from 0.8 to 0.0) while holding other parameters constant, i.e., $4Nv = 4Nv' = 0.004$, $4Nm = 1$ and $4Nc = 0.4$. This example corresponds to mutation rates $v' = v = 10^{-4}$, selection coefficients $s' = 0.02 \sim 0.00$, $c = 0.01$ and migration rate $m = 0.025$, if the effective size of each deme (N) is 10, but $v' = v = 10^{-5}$, $s' = 0.002 \sim 0.000$, $c = 0.001$ and $m = 0.0025$ if $N = 100$.

Table 1 Relationship between the index D or D_K and the mean gene frequency $\bar{x}$ at equilibrium. In this numerical example, S' or 4Ns' is changed from 0.8 to 0.0, while holding constant 4Nv = 4Nv' = 0.004, 4Nm = 1 and 4Nc = 0.4. For details, see text.

(Index)		(4Ns')	(Mean)
D	D_K	S'	$\bar{x}$
-0.40	-0.4032	0.80	0.016
-0.10	-0.1032	0.50	0.064
-0.016	-0.0192	0.416	0.263
0.0000	-0.0032	0.40	0.454
0.0032	0.0000	0.3968	0.500
0.02	0.01683	0.38	0.700
0.04	0.0368	0.36	0.815
0.07	0.0668	0.33	0.885
0.40	0.3968	0.00	0.980

In Figure 1, the relationship between D and $\bar{x}$ is illustrated by a solid line, showing very rapid transition from $\bar{x} \approx 0$ to $\bar{x} \approx 1$ as D changes from a negative to a positive value. Although D is very useful in predicting the behavior of the mean $\bar{x}$, a more accurate predictor is D_K, as confirmed by the present numerical studies (see Table 1).

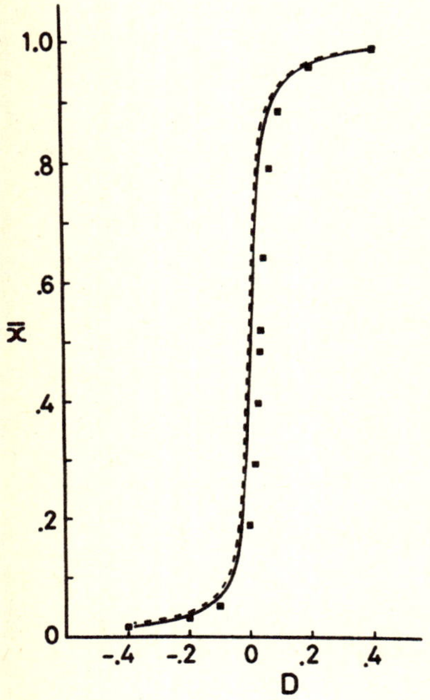

Figure 1. Relationship between the index $D = c/m - 4Ns'$ and the mean gene frequency $\bar{x}$ at equilibrium. In this figure, the solid curve represents the exact relationship obtained by numerical solution of the relevant diffusion equation, while the broken curve is the approximate one obtained by Equation (26). The square dots show the corresponding relationship derived by a discrete model involving matrix multiplication assuming very small deme size N = 2. For details, see text.

In figure 1, the broken line represents the approximate relationship between D and $\bar{x}$ as computed by using Equation (26). The agreement between the approximate values (broken curve) and the exact values (solid line) is quite good.

The diffusion model as formulated in this paper implicity assumes that the effective population size (N) of each deme is sufficiently large so that the process of change in gene frequency can be approximated by a time-continuous stochastic process. On the other hand, the biologically important situation which we have in mind corresponds to cases in which N is rather small such as N = 10 or N = 100.

In order to check if the present model is essentially valid for a small deme size, I have formulated a discrete analogue of the diffusion model and investigated numerically if this gives essentially the same result. In this discrete model, each deme consists of N diploid individuals, and relative frequency of demes having i $\underline{A}'$ alleles is denoted by F(i), where i = 0, 1, ⋯, 2N. Each generation consists of competition between demes, migration, selection within demes, and mutation (in this order), and, finally, random sampling of gametes is assumed to occur to produce demes in the next generation. Except for the process of random drift, all the changes were computed deterministically. For example, the change of frequency classes by interdemic competition follows the scheme:

$$F(i)(1 + cx)/(1 + c\bar{x}) \rightarrow F(i),$$

where $x = i/(2N)$ and $\bar{x} = \Sigma_{i=0}^{i=2N} xF(i)$. The change due to random drift was represented by multiplying $(2N + 1) \times (2N + 1)$ matrix with the elements

$$a_{ji} = \frac{(2N)!}{j!(2N - j)!} p_i^j (1 - p_i)^{2N-j},$$

where a_{ji} stands for the transition probability that a deme containing i $\underline{A}'$ alleles initially (i.e., before the occurrence of migration, selection and mutation) moves to the class containing j $\underline{A}'$ alleles after sampling. In this formula, p_i is the frequency of $\underline{A}'$ after the occurrences of migration, individual selection and mutation. More precisely, let $p_i^{(0)} = i/(2N)$, then $p_i^{(0)} + m(\bar{x} - p_i^{(0)}) \rightarrow p_i^{(1)}$ by migration; $\left[p_i^{(1)} - s'(p_i^{(1)})^2\right] / \left[1 + s'(1 - 2p_i^{(1)})\right] \rightarrow p_i^{(2)}$ by individual selection; and, $p_i^{(2)} + \left[1 - p_i^{(2)}\right]v' - p_i^{(2)}v \rightarrow p_i$ by mutation.

This process was repeated for a large number of generations until the steady state was reached with respect to the distribution, F(i), i = 0, 1, ⋯, 2N. In Figure 1, square dots show the results of such a discrete treatment assuming N = 2, v = v' = 0.0005, m = 0.125, c = 0.05 and changing the selection coefficient over the range s' = 0.1 ∿ 0.0. Note that with these parameters, we have 4Nv = 4Nv' = 0.004, 4Nm = 1, 4Nc = 0.4 and 4Ns' = 0.8 ∿ 0.0, corresponding exactly to the case of diffusion treatment as illustrated by solid curve in Figure 1. It is reassuring to note that even for such a small deme size as N = 2 (i.e., two monoecious breeding individuals per deme), the results of the diffusion equation method remain essentially valid.

Concluding remarks

In the conventional treatment of the stochastic process of gene frequencies in finite populations, it is assumed that natural selection acts through survival and reproduction of individuals (as postulated by Darwin) rather than through competition between demes. Thus, the appropriate diffusion equation for the probability distribution (ϕ) is

$$\frac{\partial \phi}{\partial t} = L(\phi),$$

where L is the standard, Kolmogorov forward differential operator commonly used in population genetics, and, in the case of single locus with a pair of alleles, it is given by (5a). On the other hand, an additional term is needed if we want to incorporate the process of group selection. Previously, Boorman and Levitt (1973) investigated group selection using the equation of the form

$$\frac{\partial \phi}{\partial t} = L(\phi) - E(x)\phi,$$

where $E(x)$, >0, is the extinction function. In this approach, all the frequency classes vanish at the limit $t = \infty$, i.e., $\phi(x, \infty) = 0$, for all x; the cohort eventually becomes extinct. This type of formulation is known as "diffusion with killing" in the theory of stochastic processes, since at each point x there is a probability $E(x)dt$ that the process is killed (see Karlin and Taylor, 1981).

The present model differs from the ordinary models which assume "killing," because in this model, even if a certain fraction of demes become extinct, some of the other demes multiply and the total number of demes remains constant in each generation. This is clearly seen by setting $f(x) = 1$ in Equation (6). In this case, $f'(x) = f''(x) = 0$ and $E\{c(x - \bar{x})f(x)\} = 0$, so that the right hand side of Equation (6) vanishes. Thus, we have

$$\frac{\partial}{\partial t} E\{1\} = \frac{\partial}{\partial t} \int_0^1 \phi(x; t)dx = 0. \tag{31}$$

Since an evolved trait is a consequence of survival rather than extinction, our model is probably more suitable than that of Boorman and Levitt (1973) to treat evolution of altruistic traits through group selection.

As shown in the previous section, the index D_K as defined by Equation (26) serves as a good indicator for predicting whether group or individual selection prevails over the other when the two counteract. Thus, if

(32) $$D_K = c/(v + v' + m) - 4Ns' > 0,$$

group selection overrides individual selection, while if $D_K < 0$, individual selection prevails.

If the mutation rates $(v + v')$ are much smaller than the migration rate (m). The above formula may be approximated by

(33) $$D = c/m - 4Ns' > 0,$$

which leads to

(34) $$4Nm < c/s'.$$

This is essentially equivalent to Aoki's (1982) condition, $nm \lesssim k/(2s)$, where n is the size of each deme, k is a measure of the intensity of group selection, m is the migration rate and s is the selection coefficient against the altruistic allele. If we take into account the fact that in Aoki's model, individuals are haploid, and therefore his n corresponds to 2N, his condition may be expressed in our terminology as $2Nm \lesssim c/(2s')$, which agrees essentially with the above condition (33). It is possible that condition (32) is more suitable, particularly when the migration rate is very low so that the mutation rates can not be neglected. It is hoped that the mathematical theory formulated in this paper will turn out to be useful for further investigation of this fascinating subject involving group selection and evolution of altruistic traits.

Contribution No. 1653 from the National Institute of Genetics, Mishima, Shizuoka-ken, 411 Japan. This work is supported in part by a Grant-in-Aid from the Ministry of Education, Science and Culture of Japan.

References

Aoki, K. (1982). A condition for group selection to prevail over counteracting individual selection. Evolution, 36, 832-842.

Boorman, S. A. & Levitt, P. R. (1973). Group selection on the boundary of a stable population. Theor. Pop. Biol., 4, 85-128.

Crow, J. F. & Kimura, M. (1970). An Introduction to Population Genetics Theory. New York: Harper & Row, reprinted (1977) by Burgess, Minneapolis, MN.

Darwin, C. (1859). The Origin of Species by Means of Natural Selection. London: John Murray.

Karlin, S. & Taylor, H. M. (1981). A Second Course in Stochastic Processes. New York: Academic Press.

Kimura, M. (1964). Diffusion models in population genetics. J. Appl. Probab., 1, 177-232.

Kimura, M. (1971). Theoretical foundation of population genetics at the molecular level. Theor. Pop. Biol., 2, 174-208.

Kimura, M. (1983a). The Neutral Theory of Molecular Evolution. Cambridge Univ. Press, Cambridge, England.

Kimura, M. (1983b). Diffusion model of intergroup selection, with special reference to evolution of an altruistic character. Proc. Natl. Acad. Sci. USA, 80, 6317-6321.

Kimura, M. & Ohta, T. (1971). Theoretical Aspects of Population Genetics. Princeton: Princeton Univ. Press.

Nei, M. (1975). Molecular Population Genetics and Evolution. Amsterdam: North-Holland.

Ohta, T. & Kimura, M. (1971). Linkage disequilibrium between two segregating nucleotide sites under the steady flux of mutations in a finite population. Genetics 68, 571-580.

Wilson, E. O. (1975). Sociobiology. The Belknap Press of Harvard Univ. Press.

Wright, S. (1931). Evolution in Mendelian populations. Genetics 16, 97-159.

Wright, S. (1938). The distribution of gene frequencies under irreversible mutation. Proc. Natl. Acad. Sci. USA, 24, 253-259.

On Laplacian Operators of Generalized Brownian Functionals

Hui-Hsiung Kuo*
Department of Mathematics
Louisiana State University
Baton Rouge, La. 70803, USA

§1. Outline of white noise calculus.

White noise calculus is the study of generalized functionals of white noise [3, 4, 5, 6, 7, 8, 9, 11, 15]. It can be regarded as an infinite dimensional calculus with the finite dimensional space $\mathbb{R}^r$, the Lebesgue measure dx and the coordinate system $(x_1, \ldots, x_r)$ being replaced by the space $\mathcal{S}^*$ of tempered distributions, the Gaussian measure $d\mu(x)$ and $\{\dot{B}(t);\ t \in \mathbb{R}\}$, respectively. The following list gives a comparison between the finite dimensional and infinite dimensional cases:

finite dimensions	infinite dimensions
$\mathbb{R}^r$	$\mathcal{S}^*$
dx	$d\mu(x)$
$\mathcal{D} \subset L^2(\mathbb{R}^r) \subset \mathcal{D}'$	$(L^2)^+ \subset L^2(\mathcal{S}^*) \subset (L^2)^-$
$(x_1, \ldots, x_r)$	$\{\dot{B}(t);\ t \in \mathbb{R}\}$
$f(x_1, \ldots, x_r)$	$\varphi(\dot{B}(t);\ t \in \mathbb{R})$
$\partial_j = \partial/\partial x_j$	$\partial_t = \partial/\partial \dot{B}(t)$
∂_j^*	∂_t^*
$x_j \cdot$	$\dot{B}(t) \cdot$
Δ	Δ_G(Gross), Δ_B(Beltrami) Δ_L(Levy), Δ_V(Volterra)

The purpose of this paper is to study various types of Laplacian operators in the white noise calculus (also called Hida calculus[18]) from the viewpoint of nonstandard analysis. We remark that there are other kinds of infinite dimensional calculus, e.g. Malliavin's calculus [10] and calculus of differentiable measures [14]. A connection between the white noise calculus and Malliavin's calculus has been described recently by Potthoff [17].

---

*Research supported by Japan Society for the Promotion of Science and NSF grant DMS-8501775.

§2. The space of generalized Brownian functionals.

Let $\mathscr{S}$ be the Schwartz space of rapidly decreasing smooth functions on $\mathbb{R}$. The dual space $\mathscr{S}^*$ of $\mathscr{S}$ consists of tempered distributions and carries the standard Gassian measure μ with the characteristic functional given by

$$\int_{\mathscr{S}^*} e^{i(x,\xi)} d\mu(x) = \exp[-\tfrac{1}{2}\|\xi\|^2], \quad \xi \in \mathscr{S},$$

where $\|\cdot\|$ denotes the $L^2(\mathbb{R})$-norm. It is well-known that $L^2(\mathscr{S}^*)$ admits the following orthogonal decomposition:

$$L^2(\mathscr{S}^*) = \sum_{n=0}^{\infty} \oplus K_n,$$

where K_n is the space of multiple Wiener integrals φ of order n, i.e. $\varphi(x) = \int_{\mathbb{R}^n} f(u_1,\ldots,u_n) dB(u_1, x)\ldots.dB(u_n, x)$. Here $B(u, x) = \langle x, 1_{(0\wedge u,\ 0\vee u]}\rangle$, $x \in \mathscr{S}^*$, is a Brownian motion.

The S-transform [11] of a Brownian functional φ in $L^2(\mathscr{S}^*)$ is defined by

$$(S\varphi)(\xi) = \int_{\mathscr{S}^*} \varphi(x+\xi) d\mu(x), \quad \xi \in \mathscr{S}.$$

If φ is the multiple Wiener integral of f, then its S-transform is given by

$$(S\varphi)(\xi) = \int_{\mathbb{R}^n} f(u_1,\ldots,u_n)\xi(u_1)\ldots\xi(u_n) du_1\ldots du_n, \quad \xi \in \mathscr{S}.$$

Note that the multiple Wiener integral φ is defined only for $f \in L^2(\mathbb{R}^n)$, while the S-transform of φ makes sense for f in $\mathscr{S}^*(\mathbb{R}^n)$. In **particular**, we can consider the generalized multiple Wiener integrals [4] of functions f in the Sobolev space $H^{-(n+1)/2}(\mathbb{R}^n)$. Thus we have the following diagram:

$$\begin{array}{ccccc} K_n^{(n)} & \subset & K_n & \subset & K_n^{(-n)} \\ \updownarrow & & \updownarrow & & \updownarrow \\ \sqrt{n!}\,\hat{H}^{(n+1)/2}(\mathbb{R}^n) & \subset & \sqrt{n!}\,\hat{L}^2(\mathbb{R}^n) & \subset & \sqrt{n!}\,\hat{H}^{-(n+1)/2}(\mathbb{R}^n). \end{array}$$

Here $\wedge$ denotes the symmetric functions and the arrow signs are the unitary isomorphisms associating φ with f. For $n=0$, $K_0^{(0)}$ is the nonstandard real number system $\mathbb{R}'$. $K_n^{(n)}$, K_n and $K_n^{(-n)}$ are regarded as vector spaces over $\mathbb{R}'$. The space $(L^2)^+$ of test Brownian functionals and the space $(L^2)^-$ of generalized Brownian functionals are defined by

$$(L^2)^+ = \sum_{n=0}^{\infty} \oplus K_n^{(n)}$$

$$(L^2)^- = \sum_{n=0}^{\infty} \oplus K_n^{(-n)}.$$

§3. U-functionals.

For a generalized Brownian functional φ in $(L^2)^-$, we define its U-functional by

$$U[\varphi](\xi) = \sum_{n=0}^{\infty} \int_{\mathbb{R}^n} f_n(u_1, \ldots, u_n)\xi(u_1)\ldots\xi(u_n)du_1\ldots du_n, \quad \xi \in \mathcal{S},$$

where $f_n \in \hat{H}^{-(n+1)/2}(\mathbb{R}^n)$ represents φ_n in the $(L^2)^-$-expansion of φ, i.e. $\varphi = \sum_{n=0}^{\infty} \varphi_n$, $\varphi_n \in K_n^{(-n)}$. It is easy to see that for φ in $L^2(\mathcal{S}^*)$, its U-functional is given by its S-transform, i.e. $U[\varphi](\xi) = (S\varphi)(\xi)$, $\xi \in \mathcal{S}$. In general, the computation of U-functionals requires the renormalization [3, 4, 5, 6, 9, 15]. A large part of white noise calculus is studied in terms of the U-functionals.

Example 1 [4]. Additive renormalization yields a generalized Brownian functional $\varphi(\dot{B}) = :\dot{B}(t)^n:$. The U-functional of φ is given by $U[\varphi](\xi) = \xi(t)^n$. For $n = 1$, $:\dot{B}(t): = \dot{B}(t)$ and the collection $\{\dot{B}(t);\ t \in \mathbb{R}\}$ is taken to be a coordinate system in the white noise calculus.

Example 2 [6, 15]. Multiplicative renormalization yields a generalized Brownian functional $\varphi(\dot{B}) = :\exp[-i\int f(u)\dot{B}(u)du]:$, $f \in \mathcal{S}^*$ and $f \notin L^2(\mathbb{R})$. Its U-functional is given by $U[\varphi](\xi) = \exp[-i\int f(u)\xi(u)du]$.

Example 3 [15]. The generalized Brownian functional $\varphi(\dot{B}) = :H_n(\dot{B}(t);((1-2c)dt)^{-1})\exp[c\int \dot{B}(u)^2du]:$, $c \neq 1/2$, comes from a mixture of additive and multiplicative renormalization. Its U-functional is given by

$$U[\varphi](\xi) = \frac{1}{n!}\left(\frac{\xi(t)}{1-2c}\right)^n \exp\left[\frac{c}{1-2c}\int \xi(u)^2du\right].$$

In fact, from the nonstandard analysis viewpoint, the space $(L^2)^-$ of generalized Brownian functionals contains not only $:\dot{B}(t)^n:$, but also $\dot{B}(t)^n$. In particular, $\dot{B}(t)^2 = :\dot{B}(t)^2: + \frac{1}{dt}$ with U-functional given by $U[\dot{B}(t)^2](\xi) = \xi(t)^2 + \frac{1}{dt}$. It is well-known that in the Ito calculus $(dB(t))^2 = dt$. But in the white noise calculus $(dB(t))^2 = dt + :\dot{B}(t)^2:(dt)^2$. The extra term $:\dot{B}(t)^2:(dt)^2$ shows the difference and is essential for the study of generalized Brownian functionals, e.g.

$$\int_a^b \dot{B}(t):\dot{B}(t)^2:(dt)^2 = 2(B(b) - B(a)).$$

§4. $\dot{B}(t)$-differentiation.

In the white noise calculus, the collection $\{\dot{B}(t);\ t \in \mathbb{R}\}$ is taken as a coordinate system. The coordinate differentiation $\partial_t \equiv \partial/\partial\dot{B}(t)$, i.e. $\dot{B}(t)$-differentiation, of a generalized Brownian functional φ is defined as follows. Suppose the U-functional U of φ has the first variation δU given by

$$(\delta U)_\xi(\eta) = \int_{\mathbb{R}} U'(\xi;\ u)\,\eta(u)du, \qquad \xi,\ \eta \in \mathcal{S},$$

and that $U'(\cdot\,;\ t)$ is a U-functional. Then we define $\partial_t\varphi$ to be the generalized Brownian functional with U-functional $U'(\cdot\,;\ t)$, i.e.

$$U[\partial_t\varphi](\xi) = U'(\xi;\ t).$$

From the nonstandard analysis point of view, the coordinate $\dot{B}(t)$-differentiation has the following property:

$$\partial_t\dot{B}(u) = \delta_t(u) = \begin{cases} \frac{1}{dt}, & t = u \\ 0, & t \neq u. \end{cases}$$

This property can be used to differentiate generalized Brownian funtionals without appealing to the U-functionals. Note that in the finite dimensional case we have $\partial_i x_j = \delta_{ij}$.

The adjoint ∂_t^* of ∂_t is defined by duality, i.e. $\langle\partial_t^*\varphi,\ \psi\rangle = \langle\varphi,\ \partial_t\psi\rangle$, $\varphi \in (L^2)^-$, $\psi \in (L^2)^+$. If U is the U-functional of φ, then the U-functional of $\partial_t^*\varphi$ is given by

$$U[\partial_t^*\varphi](\xi) = \xi(t)U(\xi).$$

The coordinate multiplication is defined by

$$\dot{B}(t)\varphi = \partial_t + \partial_t^*\varphi.$$

§5. Laplacian operators.

In the theory of abstract Wiener space [1, 12], two Laplacians and the diffusion processes associated with them have been studied [2, 13, 16]. Let (H , B) be an abstract Wiener space. The Gross Laplacian Δ_G and the Beltrami Laplacian Δ_B are defined by

$$(\Delta_G f)(x) = \mathrm{trace}_H f''(x)$$

$$(\Delta_B g)(x) = \mathrm{trace}_H g''(x) - (x,\ g'(x)),$$

where the primes denote the H-derivatives, i.e. derivatives in the directions of H. $f''(x)$ is assumed to be a trace class operator of H for each x so that $\Delta_G f$ is defined pointwise. On the other hand, for g satisfying $|g'(\cdot)|_H \in L^2(B)$ and $\|g''(\cdot)\|_{HS} \in L^2(B)$, it has been shown in

[16] that

$$\text{trace}_H Pg''(x) - (Px, g'(x))$$

converges in $L^2(B)$ as P converges strongly to the identity in the set of finite dimensional orthogonal projections of H. Thus $\Delta_B g$ is defined in the $L^2(B)$ sense.

The Gross Laplacian Δ_G is the infinitesmal generator of $\sqrt{2}\,W_t$ and W_t is a Brownian motion on (H, B) with the transition probabilities

$$p_t(x, dy) = \mu\left(\frac{dy - x}{\sqrt{t}}\right),$$

where μ is the standard Gaussian measure on (H, B). The Beltrami Laplacian Δ_B is a self-adjoint operator and its restriction to the space K_n of homogeneous chaos of order n is the multiplication by -n. It is the infinitesmal generator of Ornstein-Uhlenbeck process on (H, B) with the transition probabilities

$$q_t(x, dy) = p_{1-e^{-2t}}(e^{-t}x, dy).$$

Now, the white noise calculus is the study of generalized Brownian functionals on the white noise space $(\mathcal{S}^*, \mu)$ with $\{\dot{B}(t); t \in \mathbb{R}\}$ as a coordinate system. The support of μ is actually in a much smaller space than $\mathcal{S}^*$. Let $\mathcal{S}_p{}^*$ be the completion of $L^2(\mathbb{R})$ with respect to the following norm:

$$\|x\|_{-p} = \left[\sum_{n=0}^{\infty}(2n+1)^{-p}\langle x, \xi_n\rangle^2\right]^{1/2},$$

where ξ_n is the Hermite functions of order n and $\{\xi_n;\ n = 0, 1, 2, \ldots\}$ is an orthonormal basis for $L^2(\mathbb{R})$. Then $(L^2(\mathbb{R}), \mathcal{S}_p{}^*)$ is an abstract Wiener space for any $p > 1$. Therefore, μ is supported in $\mathcal{S}_p{}^*$, $p > 1$. It can be checked that $\delta_t \in \mathcal{S}_p{}^*$ for $p > \frac{5}{6}$ and

$$\delta_t = \sum_{n=0}^{\infty}\langle\delta_t, e_n\rangle_{-p}\, e_n,$$

where $e_n(x) = (2n+1)^{p/2}\xi_n(x)$ and $\{e_n;\ n = 0, 1, 2, \ldots\}$ is an orthonormal basis for $\mathcal{S}_p{}^*$. It is this expansion of δ_t that relates ordinary Brownian functionals in the white noise calculus to differential calculus on an abstract Wiener space.

Theorem 1. Suppose φ is a twice $L^2(\mathbb{R})$-differentiable function on some $\mathcal{S}_p{}^*$, $p > 1$, such that $\varphi'(x) \in \mathcal{S}_p$ and $\varphi''(x)$ is a trace class operator of $L^2(\mathbb{R})$ for all x in $\mathcal{S}_p{}^*$. Then φ is twice $\dot{B}(t)$-differentiable and

$$\Delta_G\varphi = \int_{\mathbb{R}} \partial_t^2\varphi\, dt, \qquad \Delta_B\varphi = -\int_{\mathbb{R}} \partial_t^*\partial_t\varphi\, dt.$$

Remark. Note that in white noise calculus, $\{\dot{B}(t);\ t\in\mathbb{R}\}$ is regarded as a coordinate system so that $\int_{\mathbb{R}}\partial_t^2\,dt$ is the infinite dimensional analogue of $\Sigma_{j=1}^r \partial_j^2$.

Proof. Under the assumption, we have

$$\partial_t\varphi(x) = (\varphi'(x),\ \delta_t).$$

By the $\mathcal{S}_p{}^*$-expansion of δ_t for $p>1$,

$$\partial_t\varphi(x) = \sum_{n=0}^{\infty}(\varphi'(x),\ e_n)\langle\delta_t,\ e_n\rangle_{-p}.$$

Therefore,

$$\partial_t^2\varphi(x) = \sum_{k,n=0}^{\infty}(\varphi''(x)e_k,\ e_n)\langle\delta_t,\ e_k\rangle_{-p}\langle\delta_t,\ e_n\rangle_{-p}.$$

Note that $e_n = (2n+1)^{p/2}\,\xi_n$ and

$$\langle\delta_t,\ e_n\rangle_{-p} = (2n+1)^{-p/2}\,\xi_n(t).$$

Hence,

$$\partial_t^2\varphi(x) = \sum_{k,n=0}^{\infty}(\varphi''(x)\xi_k,\ \xi_n)\xi_k(t)\xi_n(t).$$

Since $\{\xi_n;\ n=0,\ 1,\ 2,\ \ldots\}$ is an orthonormal basis for $L^2(\mathbb{R})$, it follows easily that

$$\int_{\mathbb{R}}\partial_t^2\varphi(x)\,dt = \sum_{n=0}^{\infty}(\varphi''(x)\xi_n,\ \xi_n)$$
$$= (\Delta_G\varphi)(x).$$

Furthermore, we have

$$(x,\ \varphi'(x)) = \int_{\mathbb{R}}x(t)\varphi'(x)(t)dt$$
$$= \int_{\mathbb{R}}(\partial_t+\partial_t^*)\partial_t\varphi(x)dt$$
$$= \int_{\mathbb{R}}\partial_t^2\varphi(x)dt + \int_{\mathbb{R}}\partial_t^*\partial_t\varphi(x)dt.$$
$$= \Delta_G\varphi(x) + \int_{\mathbb{R}}\partial_t^*\partial_t\varphi(x)dt.$$

Since $\Delta_B\varphi(x) = \Delta_G\varphi(x) - (x,\ \varphi'(x))$, we get

$$\Delta_B\varphi = -\int_{\mathbb{R}}\partial_t^*\partial_t\varphi\,dt. \qquad \#$$

For the space $(L^2)^-$ of generalized Brownian functionals, we have two new Laplacians defined as follows. Let U be the U-functional of φ in $(L^2)^-$. Suppose that the second variation of U is given by

$$(\delta^2U)_\xi(\eta,\zeta) = \int_{\mathbb{R}}U_1''(\xi;t)\eta(t)\zeta(t)dt + \iint_{\mathbb{R}^2}U_2''(\xi;t,s)\eta(t)\zeta(s)dtds.$$

If $\int_{\mathbb{R}} U_1''(\cdot;\ t)dt$ is a U-functional, then we define the Levy Laplacian $\Delta_L\varphi$ of φ to be the generalized Brownian functional whose U-functional is $\int_{\mathbb{R}} U_1''(\cdot;\ t)dt$, i.e.

$$U[\Delta_L \varphi](\xi) = \int_{\mathbb{R}} U_1''(\xi;\ t)dt.$$

If $\int_{\mathbb{R}} U_2''(\cdot;\ t,\ t)dt$ exists and is a U-functional, then we define the Volterra Laplacian $\Delta_V\varphi$ of φ to be the generalized Brownian functional whose U-functional is $\int_{\mathbb{R}} U_2''(\cdot;\ t,\ t)dt$, i.e.

$$U[\Delta_V\varphi](\xi) = \int_{\mathbb{R}} U_2''(\xi;\ t,\ t)dt.$$

By using the $\dot{B}(t)$-differentiation operator ∂_t, we can express $\Delta_L\varphi$ and $\Delta_V\varphi$ in terms of the $\dot{B}(t)$-derivatives of φ. Suppose $\partial_t^2\varphi$ is given by

$$\partial_t^2 \varphi = \varphi_1(t)\frac{1}{dt} + \varphi_2(t),$$

where $\varphi_1(t)$ and $\varphi_2(t)$ are generalized Brownian functionals such that $\varphi_1(\cdot)$ and $\varphi_2(\cdot)$ are integrable. Then

$$\Delta_L \varphi = \int_{\mathbb{R}} \varphi_1(t)dt$$

$$\Delta_V \varphi = \int_{\mathbb{R}} \varphi_2(t)dt.$$

Note that we can use nonstandard analysis notation to rewrite $\Delta_L\varphi$, i.e.

$$\Delta_L \varphi = \int_{\mathbb{R}} \partial_t^2 \varphi(dt)^2.$$

It is clear that if φ is a twice $\dot{B}(t)$-differentiable ordinary Brownian functional then $\Delta_L\varphi = 0$.

Example 1. $\varphi = \int\cdots\int_{\mathbb{R}^n} f(u_1,\ldots,u_n):\dot{B}(u_1)\ldots\dot{B}(u_n):du_1\ldots du_n$ is an ordinary Brownian functional in K_n when $f \in \hat{L}^2(\mathbb{R}^n)$. Suppose that f is continuous. Then

$$\partial_t\varphi = n\int\cdots\int_{\mathbb{R}^{n-1}} f(t,u_2,\ldots,u_n):\dot{B}(u_2)\ldots\dot{B}(u_n):du_2\ldots du_n$$

$$\partial_t^*\partial_t\varphi = n\int\cdots\int_{\mathbb{R}^{n-1}} f(t,u_2,\ldots,u_n):\dot{B}(t)\dot{B}(u_2)\ldots\dot{B}(u_n):du_2\ldots du_n$$

$$\partial_t^2 \varphi = n(n-1)\int\cdots\int_{\mathbb{R}^{n-2}} f(t,t,u_3,\ldots,u_n):\dot{B}(u_3)\ldots\dot{B}(u_n):du_3\ldots du_n.$$

The Laplacians of φ are given by:

$$\Delta_G\varphi = n(n-1)\int_{\mathbb{R}^{n-2}}\cdots\int[\int_{\mathbb{R}} f(t,t,u_3,\ldots,u_n)dt]:\dot{B}(u_3)\ldots\dot{B}(u_n):du_3\ldots du_n$$

$$\Delta_B\varphi = -n\varphi$$

$$\Delta_L\varphi = 0$$

$$\Delta_V\varphi = \Delta_G\varphi.$$

Example 2. $\varphi = \int_{\mathbb{R}} f(u):B(u)^n:du$ is a normal generalized Brownian functional in $K_n^{(-n)}$ [4, 8]. We have

$$\partial_t\varphi = nf(t):\dot{B}(t)^{n-1}:$$

$$\partial_t^*\partial_t\varphi = nf(t):\dot{B}(t)^n:$$

$$\partial_t^2\varphi = n(n-1)f(t):\dot{B}(t)^{n-2}: \frac{1}{dt}.$$

The Laplacians of φ are given by:

$\Delta_G\varphi$: does not exist

$$\Delta_B\varphi = -n\varphi$$

$$\Delta_L\varphi = n(n-1)\int_{\mathbb{R}} f(t):\dot{B}(t)^{n-2}:dt$$

$$\Delta_V\varphi = 0$$

Example 3. $\varphi = \iint_{\mathbb{R}^2} f(u, v):\dot{B}(u)^n\dot{B}(v)^k:dudv$ is also a normal generalized Brownian functional and

$$\partial_t\varphi = n\int_{\mathbb{R}} f(t, v):\dot{B}(t)^{n-1}\dot{B}(v)^k:dv$$
$$+ k\int_{\mathbb{R}} f(u, t):\dot{B}(u)^n\dot{B}(t)^{k-1}:du$$

$$\partial_t^*\partial_t\varphi = n\int_{\mathbb{R}} f(t, v):\dot{B}(t)^n\dot{B}(v)^k:dv$$
$$+ k\int_{\mathbb{R}} f(u, t):\dot{B}(u)^n\dot{B}(t)^k:du$$

$$\partial_t^2\varphi = [n(n-1)\int_{\mathbb{R}} f(t, v):\dot{B}(t)^{n-2}\dot{B}(v)^k:dv$$
$$+ k(k-1)\int_{\mathbb{R}} f(u, t):\dot{B}(u)^n\dot{B}(t)^{k-2}:du]\frac{1}{dt}$$
$$+ 2nkf(t, t):\dot{B}(t)^{n+k-1}:.$$

Therefore, the Laplacians of φ are given by:

$\Delta_G\varphi$: does not exist

$$\Delta_B \varphi = -(n+k)\varphi$$

$$\Delta_L \varphi = n(n-1)\iint_{\mathbb{R}^2} f(u, v) :\dot{B}(u)^{n-2}\dot{B}(v)^k: dudv$$

$$+ k(k-1)\iint_{\mathbb{R}^2} f(u, v) :\dot{B}(u)^{n}\dot{B}(v)^{k-2}: dudv$$

$$\Delta_V \varphi = 2nk\int_{\mathbb{R}} f(t, t) :\dot{B}(t)^{n+k-1}: dt.$$

Example 4. $\varphi = :\exp[c\int_T \dot{B}(u)^2 du]:$, $c \neq \frac{1}{2}$, is a generalized Gaussian Brownian functional. We have

$$\partial_t \varphi = \frac{2c}{1-2c}\, \partial_t^* \varphi 1_T(t)$$

$$\partial_t^* \partial_t \varphi = \frac{2c}{1-2c} (\partial_t^*)^2 \varphi 1_T(t)$$

$$\partial_t^2 \varphi = \frac{2c}{1-2c}\, \varphi\, 1_T(t) \frac{1}{dt} + \left(\frac{2c}{1-2c}\right)^2 (\partial_t^*)^2 \varphi\, 1_T(t) .$$

Therefore, the Laplacians of φ are given by:

$\Delta_G \varphi$: does not exist

$$\Delta_B \varphi = \frac{2c}{2c-1} \int_T (\partial_t^*)^2\, \varphi dt$$

$$\Delta_L \varphi = \frac{2cT}{1-2c}\, \varphi$$

$$\Delta_V \varphi = \left(\frac{2c}{1-2c}\right)^2 \int_T (\partial_t^*)^2 \varphi dt = \frac{2c}{2c-1}\, \Delta_B \varphi .$$

Note that φ is an eigenfunction of Δ_L.

The above examples suggest the following diagram which compares the action of Laplacians on the ordinary and generalized Brownian functionals:

Laplacians	Ordinary B.F.	Generalized B.F.
Δ_G	defined	does not exist
Δ_B	$\Delta_B \vert_{K_n} = -n$	$\Delta_B \vert_{K_n^{(-n)}} = -n$
Δ_L	0	defined
Δ_V	$\Delta_V = \Delta_G$	defined

References

1. Gross, L.: Abstract Wiener spaces, Proc. 5th. Berkeley Symp. Math. Stat. Prob. 2(1965), 31-42.
2. Gross, L.: Potential theory on Hilbert space, J. Func. Anal. 1(1967), 123-181.
3. Hida, T.: Analysis of Brownian Functionals, Carleton Math. Lecture Notes, no. 13, Carleton University, Ottawa, 1975.
4. Hida, T.: Generalized multiple Wiener integrals, Proc. Japan Acad. 54A(1978), 55-58.
5. Hida, T.: Brownian motion, Application of Math. vol. 11, Springer-Verlag, Heidelberg, Berlin, New York, 1980.
6. Hida, T.: The role of exponential functions in the analysis of generalized Brownian functionals, Teor. Veroyatnost i Primenen, 27 (1982), 569-573.
7. Hida, T.: Generalized Brownian functionals, Lecture Notes in Control and Information Sciences, Springer-Verlag, 49(1983), 89-95.
8. Hida, T.: Harmonic Analysis of Brownian Functionals, Lecture Notes, Kyoto University, 1983.
9. Hida, T. and Streit, L.: On quantum theory in terms of white noise, Nagoya Math. J. 68(1977), 21-34.
10. Ikeda, N. and Watanabe, S.: An introduction to Malliavin's calculus, Taniguchi Symp. Stochastic Analysis, Katata 1983, 1-52, Kinokuniya Co., Tokyo.
11. Kubo, I. and Takenaka, S.: Calculus on Gaussian white noise, I: Proc. Japan Acad. 56A(1980), 376-380, II: ibid. 56A(1980), 411-416, III: ibid. 57A(1981), 433-437, IV: ibid. 58A(1982), 186-189.
12. Kuo, H.-H.: Gaussian measures in Banach spaces, Lecture Notes in Math. vol. 463, Springer-Verlag, Heidelberg, Berlin, New York, 1975.
13. Kuo, H.-H.: Potential theory associated with Uhlenbeck-Ornstein process, J. Func. Anal. 21(1976), 63-75.
14. Kuo, H.-H.: Differential calculus for measures on Banach spaces, Lecture Notes in Math., Springer-Verlag, 644(1978), 270-285.
15. Kuo, H.-H.: Brownian functionals and applications, Acta Applicandae Mathematicae, 1(1983), 175-188.
16. Piech, M. Ann: The Ornstein-Uhlenbeck semigroup in an infinite dimensional L^2-setting, J. Func. Anal. 18(1975), 271-285.
17. Potthoff, J.: On the connection of the white-noise and Malliavin calculi, preprint.
18. Yasue, K.: The role of the Onsager-Machlup Lagrangian in the theory of stationary diffusion process, J. Math. Phys. 20(1979), 1861-1864.

Precise Estimates for the fundamental solutions to some degenerate elliptic differential equations

Shigeo Kusuoka

Let $V_i \in C_b^\infty(\mathbb{R}^N;\mathbb{R}^N)$, $i = 0,1,\ldots,d$, and consider an S.D.E.

$$X(t,x) = x + \sum_{i=1}^{d} \int_0^t V_i(X(s,x))\circ d\theta_i(s) + \int_0^t V_0(X(s,x))\, ds .$$

Then it is well known that we can take a good version of the solution of this S.D.E. such that $\{X(t,\cdot);\ t \geqq 0\}$ is a continuous process in the space of diffeomorphisms in $\mathbb{R}^N$. Let $c \in C_b^\infty(\mathbb{R}^N;\mathbb{R})$ and let $P_t^c f(x) = E[\exp(\int_0^t c(X(s,x))\, ds) f(X(t,x))]$, $f \in C_b^\infty(\mathbb{R}^N;\mathbb{R})$. Then $\{P_t^c;\ t \geqq 0\}$ is a semigroup of continouous linear operators in $C_b^\infty(\mathbb{R}^N;\mathbb{R})$. In this article, we show some precise estimates for the operators P_t^c and their density functions under certain conditions. This is a joint work with D. Stroock and the detail will be given in the forthcoming paper.

1. Presice regularity estimates for semigroups.

First we introduce the following notion.

Definition 1.1. Let E be a separable real Hilbert space and n be an integer. We say that $f \in \underline{N}_n(\mathbb{R}^N;E)$, if f is a measurable map from $(0,\infty)\times\mathbb{R}^N\times\Theta$ into E such that

(1) $f(t,\cdot,\theta):\mathbb{R}^N \to E$ is smooth, $t \in (0,\infty)$ and $\theta \in \Theta$,

(2) $f(\cdot,x,\cdot):(0,\infty)\times\Theta \to E$ is progressively measurable, $x \in \mathbb{R}^N$,

(3) $\frac{\partial^\alpha}{\partial x^\alpha} f(t,x,\cdot)$ $\underline{G}(\underline{L};E)$, and is continuous in $t \in (0,\infty)$ for any multi-index α and x $\mathbb{R}^N$, and

(4) $\sup_{0<t\leqq T} \sup_{x\in\mathbb{R}^N} \frac{1}{t^{n/2}} \| \frac{\partial^\alpha}{\partial x^\alpha} f(t,x,\theta) \|_{p;E}^{(m)} < \infty$ for any multi-index

α, any integer $m \geqq 1$, $T > 0$ and $2 \leqq p < \infty$.

Here we use the notation in [2].

Let $\underline{A} = \bigcup_{\ell=1}^{\infty} \{0,1,\ldots,d\}^{\ell}$, and set $|\alpha| = \ell$ and $\|\alpha\| = |\alpha| + \#\{i;\alpha_i=0\}$ for $\alpha = (\alpha_1,\ldots,\alpha_\ell) \in \{0,1,\ldots,d\}^{\ell}$. Set for each multi-index $\alpha \in \underline{A}$ inductively, $V_{(\alpha)} = 0$ if $\alpha = 0$, $V_{(\alpha)} = V_\alpha$ if $\alpha = 1,\ldots,d$, and $V_{(\alpha,i)} = [V_i, V_{(\alpha)}]$, $i = 0,1,\ldots,d$, $|\alpha| \geqq 1$.

Throughout this section, we will always assume the following.

(H) There is an $\ell_0 \geqq 1$ such that for any multi-index α, $\|\alpha\| > \ell_0$, there are $a_{\alpha,\beta} \in C_b^{\infty}(\mathbb{R}^N;\mathbb{R})$, $\|\beta\| \leqq \ell_0$, satisfying $V_{(\alpha)}(x) = \sum_{\|\beta\| \leqq \ell_0} a_{\alpha,\beta}(x)\cdot V_{(\beta)}(x)$ for any x $\mathbb{R}^N$.

Remark 1.2. If there is an $\ell_0 \geqq 1$, such that

$$\inf\{ (\xi,\bar{A}_{\ell_0}(1,x)\xi) \; ; \; x \in \mathbb{R}^N,\ \xi \in \mathbb{R}^N,\ |\xi| = 1 \} > 0 ,$$

then the hypothesis (H) holds.

The main theorem in this section is the following.

Theorem 1.3. For any $\Phi \in \underline{N}_n(\mathbb{R}^N;\mathbb{R})$, $n \in \mathbb{Z}$, and $\alpha \in \underline{A}$, there are Φ_α, $\Phi_\alpha' \in \underline{N}_{n-\|\alpha\|}(\mathbb{R}^N;\mathbb{R})$ such that

$$E[\Phi(t,x)V_{(\alpha)}f(X(t,x))] = E[\Phi_\alpha(t,x)f(X(t,x))] \quad \text{and}$$

$$V_{(\alpha)}(E[\Phi(t,x)f(X(t,x))]) = E[\Phi_\alpha'(t,x)f(X(t,x))]$$

for any $f \in C_b^{\infty}(\mathbb{R}^N;\mathbb{R})$, $t > 0$, $x \in \mathbb{R}^N$.

Corollary 1.4. For any $\Phi \in \underline{N}_n(\mathbb{R}^N;\mathbb{R})$, $n \in \mathbb{Z}$, and α, $\beta \in \underline{A}$, there is a $\Phi' \in \underline{N}_{n-\|\alpha\|-\|\beta\|}(\mathbb{R}^N;\mathbb{R})$ such that

$$V_{\alpha_1}\cdots V_{\alpha_m}\{E[\Phi(t,x)(V_{\beta_1}\cdots V_{\beta_k} f)(X(t,x))]\} = E[\Phi'(t,x)f(X(t,x))]$$

for any $f \in C_b^{\infty}(\mathbb{R}^N;\mathbb{R})$, $t > 0$, $x \in \mathbb{R}^N$, where $\alpha = (\alpha_1,\ldots,\alpha_m)$ and $\beta = (\beta_1,\ldots,\beta_k)$.

As an easy consequence, we have the following.

Theorem 1.5. For any α, $\beta \in \mathbb{R}^N$ and $p > 1$, there is a constant $C < \infty$

such that

$$|(V_{\alpha_1} \cdots V_{\alpha_m} P_t^c V_{\beta_1} \cdots V_{\beta_k} f)(x)| \leqq C\, t^{-(\|\alpha\|+\|\beta\|)/2}\, P_t^c(|f|^p)(x)^{1/p}$$

for any $f \in C_b^\infty(\mathbb{R}^N;\mathbb{R})$, $0 < t \leqq 1$, $x \in \mathbb{R}^N$.

Corollary 1.6. For any α, β $\underline{A}$, there is a constant $C < \infty$ such that

$$\|V_{\alpha_1} \cdots V_{\alpha_m} P_t^c V_{\beta_1} \cdots V_{\beta_k} f\|_{L^p(\mathbb{R}^N)} \leqq C\, t^{-(\|\alpha\|+\|\beta\|)/2} \|f\|_{L^p(\mathbb{R}^N)}$$

for any $f \in C_0^\infty(\mathbb{R}^N;\mathbb{R})$, $1 \leqq p \leqq \infty$ and $0 < t \leqq 1$.

Corollary 1.6 is somehow related to the results by Rothschild and Stein [4]. In the proof of Theorem 1.3, the following plays an important role.

Theorem 1.7. Let $\lambda(t,x) =$

$\inf\{\ (\xi, \tilde{A}(t,x)\xi)\ ;\ \xi \in \mathbb{R}^N,\ (\xi, \bar{A}_{\ell_0}(t,x)\xi) = 1\ \}$. Then there are $C < \infty$, $c > 0$ and $\nu > 0$ such that

$$P[\ \lambda(t,x) < \frac{1}{K}\ ] \leqq C \exp(-c K^\nu) \quad \text{for any } K > 1,\ x \in \mathbb{R}^N \text{ and}$$

$0 < t \leqq 1$. Here $\bar{A}_{\ell_0}(1,x) = \sum_{\|\alpha\| \leqq \ell_0} V_{(\alpha)}(x) \otimes V_{(\alpha)}(x)$, and

$$\tilde{A}(t,x) = \sum_{i=1}^d \int_0^t (X(s)_*^{-1} V_i)(x) \otimes (X(s)_*^{-1} V_i)(x)\, ds,$$ the modified Malliavin covariance matrix.

Theorem 1.8. $(V_{(\alpha)}(x), \tilde{A}(t,x)^{-1} V_{(\beta)}(x)) \in \underline{N}_{-(\|\alpha\|+\|\beta\|)}(\mathbb{R}^N;\mathbb{R})$ for any $\alpha, \beta \in \underline{A}$ with $\|\alpha\|, \|\beta\| \leqq \ell_0$.

2. Precise estimates for fundamental solutions.

In this section, we assume that there are $a_i \in C_b^\infty(\mathbb{R}^N;\mathbb{R})$ such that $V_0 = \sum_{i=1}^d a_i V_i$. We also assume the following.

(H') There is an $\ell_0 \geqq 1$ such that

$$\inf\{ (\xi,\bar{A}_{\ell_0}(1,x)\xi);\ x \in \mathbb{R}^N,\ \xi \in \mathbb{R}^N,\ \|\xi\| = 1 \} > 0.$$

Then P_t^c, $t > 0$, has a smooth kernel $p^c(t,\cdot,\cdot)$.

Let $H = \{ h:[0,\infty)\to\mathbb{R}^d;\ h(0) = 0,\ h(t)$ is absolutely continuous in t and $\|h\|_H^2 = \int_0^\infty |\frac{d}{dt} h(t)|^2\, dt < \infty \}$. For each $h = (h_1,\dots,h_d)$ H, let us think of O.D.E. as follows:

$$\frac{d}{dt}\bar{X}(t,x;h) = \sum_{i=1}^{d} V_i(X(t,x;h))\frac{d}{dt}h_i(t)\ ,\ t > 0,$$

$$\bar{X}(0,x;h) = x\ .$$

Now let us define $d:\mathbb{R}^N\times\mathbb{R}^N \to [0,\infty)$ and $B(\varepsilon,x)$ $\mathbb{R}^N$, $\varepsilon > 0$, x $\mathbb{R}^N$, by

$$d(x,y) = \inf\{ \|h\|_H\ ;\ \bar{X}(1,x;h) = y \} \quad \text{and}$$

$$B(\varepsilon,x) = \{ y \in \mathbb{R}^N;\ d(x,y) < \varepsilon \}\ .$$

Our theorem is the following.

<u>Theorem 2.1.</u> There exists a constant $M > 0$ such that

$$\frac{1}{M}\,|B(\sqrt{t},x)|^{-1}\exp(\,-\,M\,\frac{d(x,y)^2}{t}\,) \leqq p^c(t,x,y) \leqq M\,|B(\sqrt{t},x)|^{-1}\exp(\,-\,\frac{d(x,y)^2}{Mt}\,)$$

for any $x,y \in \mathbb{R}^N$ and $0 < t \leqq 1$.

Combining this with Theorem 1.5, we have the following.

<u>Theorem 2.2.</u> For any n, $m \geqq 0$ and $i_1,\dots,i_{n+m} = 1,\dots,d$, there is a constant $M > 0$ such that

$$|\, V_{i_1,x}\cdots V_{i_n,x}V_{i_{n+1},y}\cdots V_{i_{n+m},y}p^c(t,x,y)\,| \leqq \frac{M}{t^{(n+m)/2}}\,|B(\sqrt{t},x)|^{-1}\exp(\,-\,\frac{d(x,y)^2}{Mt}\,)$$

for any $x,y \in \mathbb{R}^N$ and $0 < t \leqq 1$.

These results have been proved by Sanchez-Calle[5] and Jerison and Sanchez-Calle[1]. However, our approach is quite different from theirs. Our main tools are stochastic Taylor expansion and

Malliavin calculus.

References.

[1] Jerison, D., and A. Sanchez-Calle, Estimates for the heat kernel for a sum of squares of vector fields, Preprint.

[2] Kusuoka, S., and D. Stroock, Applications of Malliavin calculus I, Stochastic Analysis, Proc. Taniguchi Internatl. Symp. Katata and Kyoto, 1982, ed. by K. Ito, pp. 271-306, Kinokuniya, Tokyo.

[3] Kusuoka, S., and D. Stroock, Applications of Malliavin calculus II, J. Fac. Sci. Univ. Tokyo Sc. IA, 32(1985), 1-76.

[4]Rothschild, L.P., and E.M. Stein, Hypoelliptic differential operators and nilpotent Lie groups, Acta Math. 137(1977), 247- 320.

[5] Sanchez-Calle, A., Fundamental solution and geometry of the sum of squares of vector field, Invent. Math. 78(1984), 143-160.

Department of Mathematics
Faculty of Science
University of Tokyo
Hongo, Tokyo, JAPAN

ON STOCHASTIC ALGORITHMS IN ADAPTIVE FILTERING

Michel Métivier
Ecole Polytechnique, Palaiseau - France

The use of adaptive filtering in identification or tracking problems has introduced new classes of stochastic algorithms, the study of which raises a wide set of new problems. To deal with them it is important to recognize their deep analogy with problems related to the behaviour of randomly perturbated deterministic systems, when the perturbation depends on small parameters, and to the asymptotic stability of differential equations.

This lecture intends to give

1° - a short account of some main ideas of the "classical" theory of stochastic algorithms,

2° - an introduction to the problems as they come out from adaptive filtering and identification of systems,

3° - a presentation of some results which are essentially due to a joint work with P. Priouret and for the details of which the reader is referred to [17] or [18] or [20]. However in this last part we give the complete proof of some propositions having a general interest and which, in the form they are given here, cannot be found in the given references.

The text of this lecture was written while the author was a guest of the Japanese Society for the Promotion of Sciences (J.S.P.S.) and the University of Nagoya. It is a pleasure for him to thank these institutions for their help and hospitality.

I. Some look backward at the "classical stochastic algorithms".

1.1. Where stochastic algorithms come from.

Let us consider the following problem: Let $(X_\theta)_{\theta\in\Theta}$ be a family of random variables, μ_θ being the law of X_θ. For a given function φ one sets $\Phi(\theta) = E\varphi(X_\theta)$ and one wants to solve the equation

$$\Phi(\theta) = 0. \qquad (1.1.1)$$

If the function Φ were known one could think of using an iterative procedure of the type

$$\theta_{n+1} = \theta_n - \gamma_{n+1}\Phi(\theta_n) \tag{1.1.2}$$

to solve (1), where (γ_n) is a sequence of positive numbers.

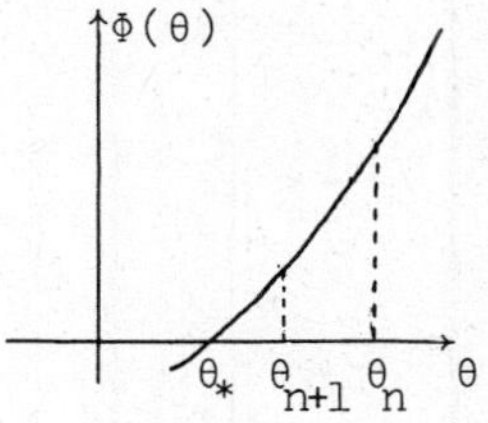

The Newton method for solving (1) is of this type. Here qualitative assumptions (like monotony) will insure the convergence of (θ_n) given by (2) to the solution θ_* of (1).

In many practical situations one wants to get an approximation of θ_* in the absence of knowledge of the probability laws p_θ, and do it from experimental observations of the variables X_θ. Suppose in particular, that for each given value θ_n of θ one can perform an experiment the outcome of which is a random variable X_n with law p_{θ_n}. One is led to think of the following stochastic substitute for (1.1.2).

$$\theta_{n+1} = \theta_n - \gamma_{n+1}\varphi(X_n) \tag{1.1.3}$$

This was the idea initially proposed by Robbins and Monro. In this setting the experiment X_n at stage n is supposed to be made "independently of the preceding ones given θ_n", which means that the conditional law of X_n given the previous experiments is p_{θ_n}. The basic motivation of such a procedure is the hope that some "natural large number averaging principle" will compensate the substitution of $\varphi(X_{\theta_n})$ to $E(\varphi(X_{\theta_n}))$ in the algorithm (1.1.2), for a proper choice of the sequence $(\gamma_n)_{n\geq 0}$.

Procedures of this type are widely used in practical statistical problems.

1.2. <u>Some typical examples</u>.

A. <u>Iterative form of the law of large numbers</u>.

Suppose that $X_\theta = X - \theta$, X being a variable with unknown law. One wants to solve $E(X) = \theta = 0$ by only observing independent realisations $X_1 \cdots X_n \cdots$ of X. The algorithm (1.1.3) with $\gamma_n = \frac{1}{n}$, takes the form

$$\theta_{n+1} = \theta_n + \frac{1}{n+1}(X_{n+1} - \theta_n) \qquad (1.2.1)$$

which converges to $\theta_* = E(X)$ since

$$\theta_{n+1} = \frac{1}{n+1}(X_1 + \cdots + X_n).$$

B. Stochastic gradient-methods.

The problem here is to solve the optimisation problem

$$\min_\theta \Phi(\theta) = \Phi(\theta_*) \quad \text{where} \quad \Phi(\theta) = E\varphi(X,\theta).$$

The stochastic gradient-methods replace a deterministic gradient algorithm of the type

$$\theta_{n+1} = \theta_n - \gamma_{n+1}\nabla_\theta \Phi(\theta_n)$$

by its stochastic analog

$$\theta_{n+1} = \theta_n - \gamma_{n+1}\nabla_\theta \varphi(X_{n+1},\theta). \qquad (1.2.2)$$

Approximate gradient algorithms have also been proposed and studied. This is the case of the Kiefer-Wolfowitz algorithm

$$\theta_{n+1} = \theta_n - \gamma_{n+1} \frac{1}{2c_n}(\varphi(X^1_{n+1}) - \varphi(X^2_{n+1})) \qquad (1.2.3)$$

where $c_n \downarrow 0$ and the couple (X^1_{n+1}, X^2_{n+1}) is the outcome of an experiment of conditional law $p_{\theta_n + c_n} \otimes p_{\theta_n - c_n}$ given $\theta_0 \cdots \theta_n$, $X^1_0 \cdots X^1_n$, X^2_n.

The algorithm (1.2.3) under suitable hypotheses behave asymptotically as the deterministic algorithm

$$\theta_{n+1} = \theta_n - \gamma_{n+1}\nabla_\theta \Phi(\theta_n)$$

where

$$\Phi(\theta) = \int \varphi(x) p_\theta(dx)$$

and converges to the solution of the minimum problem $\min_\theta \Phi(\theta)$.

1.3. Main features of the classical theory - Martingale methods.

All the preceding stochastic algorithms can be written

[A] $\theta_{n+1} = \theta_n + \gamma_{n+1} f(\theta_n, X_{n+1}) \quad \theta_n \in \mathbb{R}^d, \quad X_n \in \mathbb{R}^K.$

The huge literature devoted to them (well known references are [26],[7],[22]) essentially deals with the case which we call "Robbins-Monro" and which is characterized by the following hypothesis.

[R.M] X_{n+1} is conditionally independent of $X_0 \cdots X_n$, $\theta_0 \cdots \theta_{n-1}$ given θ_n with conditional law $\mu_{\theta_n}(dx)$.

The [R.M] hypotheses allows directly the use of martingale methods to study the asymptotic behaviour of [A]. (See [8]. For a systematic exposition of this point of view see the book by Heyde and Hall [9] or [21]). To give a rapid idea of this, let us observe that, if we set (assuming the existence of the integrals involved):

$$h(\theta) = \int f(\theta,x)\mu_\theta(dx) \tag{1.3.1}$$

the algorithm [A] can be written as

$$\theta_{n+1} = \theta_n + \gamma_{n+1} h(\theta_n) + \gamma_{n+1} m_{n+1} \tag{1.3.2}$$

where

$$m_{n+1} := f(\theta_n, X_{n+1}) - h(\theta_n)$$

is a martingale increment (consequence of [R.M]). A typical theorem of the classical theory of stochastic algorithms in the following

Theorem 1. *Let us assume that* [R.M] *holds and also:*

$$S_n^2(\theta) := \int |f(\theta,x)|^2 \mu_\theta(dx) \leq K(1 + |\theta|^2) \quad \text{for some constant } K. \tag{1.3.3}$$

We suppose that there exists $\theta_* \in \mathbb{R}^d$ *such that for all* $\varepsilon > 0$

$$\sup_{\varepsilon \leq |\theta - \theta_*| \leq \frac{1}{\varepsilon}} (\theta - \theta_*).h(\theta) < 0. \tag{1.3.4}$$

If $\sum_n \gamma_n = +\infty$ *and* $\sum_n \gamma_n^2 < \infty$, *the algorithm* [A] *converges a.s. to* θ_*.

Proof. The proof can be summarized as follows.

Let us set $T_n := \theta_n - \theta_*$. We call F_n the σ-algebra generated by $X_0 \cdots X_n$. The [R.M] property gives

$$E(|T_{n+1}^2| \mid F_n) = |T_n|^2 + 2\gamma_{n+1} T_n \cdot h(\theta_n) + \gamma_{n+1}^2 S_{n+1}^2(\theta_n). \tag{1.3.5}$$

Property (1.3.4) and (1.3.3) lead immediately to an inequality of the type

$$E(|T_{n+1}|^2) \leq E(|T_n|^2)(1 + \tilde{K}\gamma_{n+1}^2) + \tilde{K}\gamma_{n+1}^2.$$

The convergence of the series $\sum_n \gamma_n^2$ then implies

$$\sup_n E|T_n|^2 < \infty. \tag{1.3.6}$$

We then decompose (Doob-decomposition!)

$$|T_n|^2 = |T_0|^2 + N_n + B_n$$

with

$$N_n := \sum_{1 \leq k \leq n} (|T_n|^2 - E(|T_n|^2 \mid F_{n-1}) \quad \text{(martingale)}$$

$$B_n := \sum_{1 \leq k \leq n} E(|T_n|^2 \mid F_{n-1}) - |T_{n-1}|^2.$$

From (1.3.4) and (1.3.5) one easily derives

$$E[(B_n - B_{n+1})^+] \leq \tilde{C}\gamma_n^2(1 + \sup_n E|T_n|^2)$$

and therefore

$$\sum_{k \geq 1} E|E(|T_k|^2 - |T_{k-1}|^2 | F_{k-1})| = \sum_{k \geq 1} E|B_k - B_{k-1}| < \infty.$$

This shows that $(|T_n|^2)_{n \geq 0}$ is a quasi-martingale (In the sense of S. Orey. See [21] Chapter 2) and by a classical convergence theorem $\lim_{n \to \infty} |T_n|^2$ exists almost surely. We have only to prove now that

$$P(\{\lim_{n \to \infty} |T_n|^2 > 0\}) = 0.$$

But using (1.3.4) we see that on $A := \{\lim_{n \to \infty} |T_n|^2 > 0\}$ one has

$$\liminf_{n \to \infty} - T_n \cdot h(\theta_n) > 0. \tag{1.3.7}$$

Since

$$-2\gamma_{n+1} T_n \cdot h(\theta_n) \leq |B_{n+1} - B_n| + \tilde{K}\gamma_{n+1}^2(1 + |T_n|^2)$$

then

$$\sum_{n \geq 1} E(-2\gamma_{n+1} T_n \cdot h(\theta_n))^+ < \infty$$

and therefore

$$\sum_{n\geq 1} (-2\gamma_{n+1} T_n \cdot h_{n+1}(\theta_n))^+ < \infty \qquad \text{a.s.}$$

This with (1.3.7) and $\sum_{n\geq 1} \gamma_n = +\infty$ implies $P(A) = 0$.

1.4. Some basic facts and ideas to underline.

We conclude this short review of the classical theory by the following remarks.

A - The [R.M] property makes possible (see formula (1.3.1)) the writting of the algorithm in the form

$$\theta_{n+1} = \theta_n + \gamma_{n+1} h(\theta_n) + m_n \qquad (1.4.1)$$

where $M_n := \sum_{k\leq n} m_k$ is a martingale. This decomposition of the process (θ_n) allowed a direct study of the convergence of θ_n. But it can also be observed that the algorithm thus appears as a *random perturbation* of the deterministic algorithm

$$\overline{\theta}_{n+1} = \overline{\theta}_n + \gamma_{n+1} h(\overline{\theta}_n). \qquad (1.4.2)$$

In view of the Doob inequality, the perturbation (m_k) is such that

$$E|\sup_{k\geq n} \sum_{i=n+1}^{k} m_i|^2 \leq 4(\sum_{i\geq n+1} E|m_i|^2) \leq 4K(\sum_{i\geq n+1} \gamma_i^2).$$

The assumption $\sum_i \gamma_i^2 < \infty$ thus expresses that in a strong sense the perturbation tends to zero when time evolves. The asymptotic behaviour of (θ_n) can therefore be compared with the asymptotic behaviour of the deterministic algorithm $(\overline{\theta}_n)$.

B - The hypotheses (1.3.4) of Theorem 1 can be interpreted as a "stability hypothesis" for the deterministic algorithm (1.4.2). It expresses that at point θ the vector field $h(\theta)$ is directed toward the interior of the circle of center θ_* containing θ.

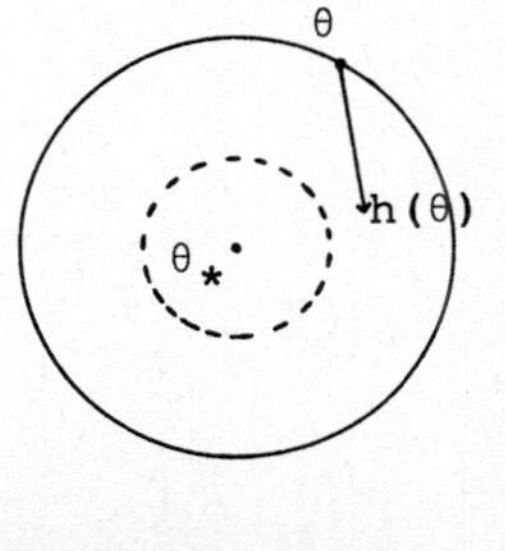

This has clearly to see with the asymptotic behaviour of the deterministic algorithm $\overline{\theta}_n$, since it says that the function $|\theta - \theta_*|^2$ is a Lyapunov function for the differential equation

$$\frac{d\overline{\theta}(t)}{dt} = h(\overline{\theta}(t)), \tag{1.4.3}$$

of which (1.4.2) is a discretized form.

C. <u>The two aspects of the problem</u>.

These considerations show that the proof of Theorem 1 could have been decomposed (in a rather lengthier but instructive way) into the proof of two successive propositions:

A first proposition using only the [R.M] property and the hypotheses of moments to obtain a majoration of

$$E(\sup_{k \geq N} |\theta_k - \overline{\theta}_{k,N}|^2),$$

where $\overline{\theta}_{k,N}$ is for $k \geq N$ the algorithm (1.4.2) with initial condition $\overline{\theta}_{N,N} = \theta_N$.

A second proposition using the stability hypotheses (1.3.4) denoted to the asymptotic properties.

This is the path followed for example in the first chapter of the lecture notes by H. Kushner and D. Clark ([12]) to extend the study when the [R.M] property is not present and which has inspired most of the modern studies which we shall go to after the next final remark.

D. <u>Stochastic algorithms and Freidlin "averaging principle"</u>.

Let us consider to simplify the case $\gamma_n = \gamma$ independent of n, γ being "small"

$$\theta_{n+1} = \theta_n + \gamma f(\theta_n, X_{n+1}).$$

We see that θ_n has "slow variation" compared with X. If we express this using a fictitious time:

$$t_n := n\gamma,$$

the algorithm becomes

$$\theta_{t_{n+1}} = \theta_{t_n} + (t_{n+1} - t_n) f(\theta_{t_n}, X_{t_n/\gamma}).$$

This situation is essentially of the same nature as the situation encountered in differential systems with a small parameter: (See [27].)

$$\begin{cases} \dfrac{d\theta^{\gamma}}{dt}(t) = f(\theta(t),\ Y(t/\gamma)) \\ Y(u) \quad \text{is a Markov or a stationary mixing process} \end{cases} \tag{1.4.4}$$

The Freidlin averaging-principle says that if we consider the "averaged equation":

$$\frac{d\overline{\theta}(t)}{dt} = h(\overline{\theta}(t))$$

where $h(\theta)$ is the average of $f(\theta, Y_u)$ for the stationary law of Y_u one may expect the following type of results:

$$\lim_{\gamma\to 0} P\{\sup_{t\leq T} |\theta^{\gamma}(t) - \overline{\theta}(t)| > \eta\} = 0$$

(Law of large numbers)

$$\lim_{\gamma\to 0} \text{law of } \left(\frac{\theta^{\gamma}(t)-\overline{\theta}(t)}{\sqrt{\gamma}}\right)_{t\geq 0} = \text{law of a gaussian diffusion}$$

(Gaussian approximation)

without speaking of large deviation estimates.

The averated equation in the Robbins-Monro case was already introduced in the form (1.4.3) or its discretized version (1.4.2).

It is clear now that dealing with general algorithm will necessitate as first step, the establishing of some "averaging principle".

II. New problem coming from adaptive filtering and identification of systems.

2.1. Example of the "adaptive equalizer".

The engineers call equalizer a linear filter with parameter $\theta \in \mathbb{R}^d$ which transforms a vector input signal U into the scalar output $\theta.U$. It is called adaptive if θ can be changed, in particular when the value of θ is a function of the state of the system in which the filter is inserted. Here is an example

A - *Learning period in a communication system.*

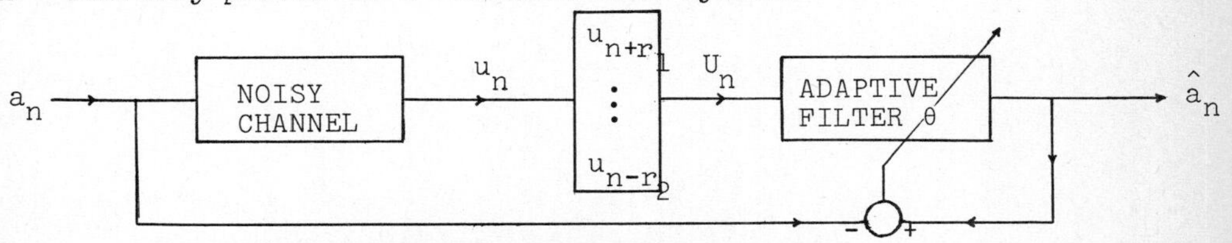

Assume that emitted signals (a_n) produce a sequence (u_n) at the output of a transmission system. For example

$$u_n = \sum_{k=0}^{N} s_k a_{n-k} + w_n \qquad (2.1.1)$$

the (s_k) being unkhown parameters of the system and w_n being a stationary but unspecified noise. The u_n are stocked, giving birth to a sequence of vectors, where r_1 and r_2 are given numbers.

$$U_n := \begin{pmatrix} u_{n+r_1} \\ \vdots \\ u_{n-r_2} \end{pmatrix}$$

The estimate $\hat{a}_n$ of the signal a_n given by the filter with parameter θ is

$$\hat{a}_n := \theta.U_n.$$

The "best" filter for the quadratic error is the solution θ_* of the minimum problem $E|\theta_*.U_n-a_n{}^2 = \min_\theta E|\theta.U_n-a_n|^2$. (In case where $w_n = 0$, θ_* would be the solution of the "deconvolution problem" associated with (2.1.1).) The ignorance of the law of U_n leads to the use of the following empirical way of approximating θ_*. During a "learning period" the transmitted sequence of signals (a_n) is available to the receiver which can compare a_n and $\hat{a}_n$. Using the ideas expressed at the beginning one may therefore think of using the following stochastic-gradient-procedure to estimate θ_* recursively:

$$\theta_{n+1} = \theta_n - \gamma_{n+1}\nabla_\theta|U_n.\theta_n - a_n|^2$$

in other words

$$\theta_{n+1} = \theta_n - \gamma_{n+1}(\theta_n.U_n - a_n)U_n. \qquad (2.1.2)$$

B - *The "blind" equalizer.*

Let us assume that we start now with a value θ_0 of θ close to the optimal value θ_* (For example after a learning period, assuming it is proved that (2.1.2) converges to θ_*). If the a_n are not available, but if we know that the a_n are digits equals to -1 or $+1$ the estimate $c_n := \text{sign}(\theta.U_n)$ will be equal to a_n with a high probability. This leads to think that the following algorithm

$$\theta_{n+1} = \theta_n - \gamma_{n+1}(\theta_n.U_n - \text{sign}\ \theta_n.U_n)U_n \qquad (2.1.3)$$

will behave approximately like (2.1.2) as long as θ_n stays in a small neighborhood of θ_*.

Actually it is used by engineers in particular to adjust the value of θ if the parameters of the transmission channel are likely to have small fluctuations after the learning period.

2.2. Main new features of these examples.

The algorithms (2.1.2) and (2.1.3) are of the same general form as those considered in Section I:

$$\theta_{n+1} = \theta_n + \gamma_{n+1} f(\theta_n, X_{n+1})$$

where (X_{n+1}) is an observed process. But there is a fundamental first defference: the "Robbins-Monro"-property [R.M] is not anymore present.

Moreover the second example shows that one needs a theory which tolerates some kinds of discontinuities for the adaptation function of $f(\theta,x)$. The extensions of the classical results, using for example hypotheses of Lipschitz type on f (see [24] or [25]) do not apply to the most current situation in adaptive filtering.

2.3. Examples in the theory of recursive identification.

We refer to the book of L. Ljung and T. Söderström [16] for numerous examples of iterative procedures leading to stochastic algorithms for the estimation of the parameters of a system. Let us just mention that if $\Lambda = (a_1,\dots,a_k, b_1,\dots,b_m, c_1,\dots,c_r)$ is the unknown parameter of a linear dynamic model of the type

$$y_n = \sum_{i=1}^{k} a_i y_{n-i} + \sum_{i=1}^{m} b_i u_{n-i} + \sum_{i=1}^{r} c_i e_{n-i} + e_n, \qquad (2.3.1)$$

where (y_n) is an observable process, the u_n's are known controls and (e_n) is a white noise, many recursive estimation procedures take the form

$$\begin{cases} \Lambda_{n+1} = \Lambda_n + \gamma_{n+1} f_n^1(\Lambda_n, R_n, Y_{n+1}) \\ R_{n+1} = R_n + \gamma_{n+1} f_n^2(\Lambda_n, R_n, Y_{n+1}), \end{cases}$$

where Y_n is a d-dimensional ovservation vector based on the vectors y_{n-i}, u_{n-i} and R_n is a positive definite symmetric $d\times d$ matrix.

A typical simple example of this is given by the "recursive form of least square estimates" (see [15] Chapter 2) in the case $r = 0$, which leads to the following algorithm, where in matrix notation;

$$Y_n := (-y_{n-1}, \dots, -y_{n-k}, u_{n-1}, \dots, u_{n-m})^T$$

and

$$P_n := \left(\frac{1}{n} \sum_{i=1}^{n} Y_{i-1} Y_{i-1}^T\right)^{-1} :$$

$$\begin{cases} \Lambda_{n+1} = \Lambda_n + \frac{1}{n+1} P_n Y_n (y_n - Y_n^T \Lambda_n)(1 + Y_n^T P_n Y_n)^{-1} \\ P_{n+1} = \frac{n+1}{n} P_n - \frac{n+1}{n^2} (P_n Y_n Y_n^T P_n)(I + \frac{1}{n} Y_n^T P_n Y_n)^{-1}. \end{cases} \tag{2.3.2}$$

Let us observe that the second equation in (2.3.2) can be written

$$P_{n+1} = P_n - \frac{1}{n+1} P_n Y_n Y_n^T P_n + \frac{1}{(n+1)^2} \Psi(n, Y_n, P_n),$$

where the matrix-valued Ψ is such that

$$\|\Psi(n, Y_n, P_n)\| \leq K(\|P_n\|)(1 + \|Y_n\|^2).$$

The algorithm (2.3.2) has thus the general form

$$\theta_{n+1} = \theta_n + \gamma_{n+1} f(\theta_n, X_n) + \gamma_{n+1}^2 \tilde{f}(n, \theta_n, X_n),$$

where f is independent of n and

$$\|f(\theta, x)\| + \|\tilde{f}(n, \theta, x)\| \leq K(\theta)(1 + |x|^\rho)$$

for some positive continuous function $K(\theta)$ and constant $\rho \geq 0$.

III. A general set up to study stochastic algorithms.

We now propose a general set up to study the stochastic algorithms mentionned in the preceding sections. The theory which we intend to present briefly is developed in several papers ([17], [18], [19]) and lecture notes ([20]). An analogous framework is considered in ([14]).

3.1. Markovian hypotheses.

The system we consider is the following

$$[A] \quad \begin{cases} \theta_{n+1} = \theta_n + \gamma_{n+1} f(n, \theta_n, Y_{n+1}) & \theta_n \in \mathbb{R}^d \\ (\theta_n, Y_n)_{n\ 0} \text{ is a Markov chain} & Y_n \in \mathbb{R}^K \end{cases}$$

Denoting by F_n the σ-algebra generated by $\theta_0,\cdots,\theta_n$, $y_0,\cdots,y_n$ we assume

[H.1] The conditional law of Y_{n+1} given F_n is given by

$$P[Y_{n+1} \in B|F_n] = \Pi_{\theta_n}(Y_n,B)$$

where $\Pi_\theta(y,dz)$ is a transition probability from $\mathbb{R}^d \times \mathbb{R}^K$ into $\mathbb{R}^K$.

[H.2] $(\gamma_n)_{n\geq 0}$ is a monotone, non increasing sequence of real numbers with $\sum_n \gamma_n = +\infty$.

Denoting as usually by $\Pi_\theta g$ the function $\Pi_\theta g(y) := \int g(z)\Pi_\theta(y,dz)$ and by Π_θ^n the iterates of the operator Π_θ (for θ fixed), we make the following hypotheses on Π_θ:

[Π.1] There exist $r \in \mathbb{N}$ and $p \geq 2$ and for every compact $K \subset \mathbb{R}^d$ some constants $\rho(Q) < 1$ and $M(Q)$ such that for all $y \in \mathbb{R}^K$

$$\sup_{\theta\in Q} \int \Pi_\theta^r(y,dz)|z|^q \leq \rho(Q)|y|^q + M(Q).$$

For every $q \geq 0$ we introduce the following "Lipschitz-coefficients" of a function g on $\mathbb{R}^K$(these coefficients are finite in particular when g has first and second derivatives with polynomial increase of order $\leq q$):

$$[g]_q := \sup_{y\neq y'} \frac{|g(y)-g(y')|}{|y-y'|(1+|y|^q+|y'|^q)}$$

$$[\![g]\!]_q := \sup_{\substack{y\neq y'\\ w\neq 0}} \frac{g(y+w)-g(y)-g(y'+w)+g(y')}{|w||y-y'|(1+|w|^q+|y|^q+|y'|^q)} .$$

With these notations we formulate the following assumptions:

[Π.2] For every compact $Q \subset G$ there exists $K(Q)$ such that for all $q \leq p-1$ and all g on $\mathbb{R}^K$

$$[\Pi_\theta g]_q \leq K(Q)[g]_q.$$

[Π.3] For every compact $Q \subset G$ there exists $K(Q)$ such that for all $q \leq p-1$ all g on $\mathbb{R}^K$, every $\theta,\theta' \in Q$ and $y \in \mathbb{R}^K$:

$$|\Pi_\theta g(y) - \Pi_{\theta'} g(y)| \leq K(Q)[g]_q|\theta-\theta'|(1 + |y|^{q+1}).$$

Example. A typical situation encountered in applications is the following:

$$Y_{n+1} = A(\theta_n)Y_n + B(\theta_n)W_{n+1}$$

where the random variables W_n are i.i.d. and where the matrix valued functions A and B are assumed to be locally lipschitzian. If one observes that

$$\Pi_\theta^n g(y) = E\{g(A^n(\theta)y + \sum_{k=0}^{n} A^k(\theta)B(\theta)W_{n-k})\}$$

one checks easily the properties [Π.2] and [Π.3].

If one also assumes as in L. Ljung ([15]) that, for every compact $Q \subset \mathbb{R}^d$ the spectrum of the matrices $A(\theta)$, $\theta \in Q$ is included in a disc with radius $r(Q) < 1$, these $\|A^n(\theta)\| \leq C(Q)r^n(Q)$ for some constant $C(Q)$. This implies immediately [Π.1].

3.2. Hypotheses on the adaption function and "averaging hypotheses".

We assume (see the example at the end of Section 2) that

[F] (i) f can be written as

$$f(n,\theta,y) = H(\theta,y) + \gamma_n^\nu \tilde{H}(n,\theta,y) \qquad \nu \geq \frac{1}{2}.$$

(ii) There exists $p \geq 0$ and for every compact $Q \subset \mathbb{R}^d$ a constant $K_1(Q)$ such that for all $y \in \mathbb{R}^K$ and $n \geq 0$

$$\sup_{\theta \in Q}|H(\theta,y)| + \sup_{\theta \in Q}|\tilde{H}(n,\theta,y)| \leq K_1(Q)(1 + |y|^p).$$

Finally we make the following "averaging hypotheses on H".

[H.3] For every $\theta \in \mathbb{R}^d$, the Markov chain $(Y_n^\theta)_{n\geq 0}$ with probability transition Π_θ has an invariant probability μ_θ such that the integral

$$h(\theta) := \int H(\theta,y)\mu_\theta(dy)$$

exists and is locally lipschitzian as a function of θ. We moreover assume that for every θ there exists a solution v_θ to the following Poisson-equation:

(i) $\qquad (I - \Pi_\theta)v_\theta(y) = H(\theta,y) - h(\theta)$

with

(ii) $\qquad \sup_{\theta \in Q}|v_\theta(y)| \leq K_2(Q)(1 + |y|^a)$ for some constants $K_2(Q)$, and $a \geq 0$.

(iii) For every Q compact in $\mathbb{R}^d$ there is a constant $K_3(Q)$ such that

$$|\Pi_\theta v_\theta(y) - \Pi_{\theta'} v_{\theta'}(y)| \leq K_3(Q)|\theta-\theta'|(1 + |y|^b) \quad \text{for all } \theta,\theta' \in Q.$$

[H.3] allows to introduce the average ordinary differential equation ([O.D.E.]). This is the equation

$$\frac{d\overline{\theta}(t)}{dt} = h(\overline{\theta}(t)).$$

3.3. The algorithm in finite time.

This is the first part of our program as announced in 1.4.D. Let us introduce the following notations:

$t_n := \gamma_1 + \cdots + \gamma_n$

$A(t) := \max\{t_n : t_n \leq t\}$

$m(n,T) := \max\{k : k \geq n,\ \gamma_{n+1} + \cdots + \gamma_{n+k} \leq T\}$

$\theta(t)$ will denote the following step-interpolation of θ_n

$\theta(t) := \theta_n$ if $t_n \leq t < t_{n+1}$

$\overline{\theta}(t,x)$ is the solution of the [O.D.E.] (8.2.1) with $\overline{\theta}(0,x) = x$ and define : $Q(x) := \inf\{t : \overline{\theta}(x,t) \notin Q\}$.

We may then state the

Theorem 2. *Let us assume all the hypotheses in (3.1) and (3.2), set* $s := \max(a,b+\rho)$ *and suppose* $p \geq 2s$. *Let* Q *be a compact in* $\mathbb{R}^d$, *let* q *be such that* $2 \leq q \leq p/2s$ *and* T *be given. Then there exists* $\eta_0 > 0$ *and a constant* $C(q,Q,T)$ *(independent of the sequence* (γ_n)*) such that for all* $x \in Q$, $y \in \mathbb{R}$, $T < s_Q(x)$ *and* $\eta < \eta_0$ *we have*

$$P_{x,y}\{\sup_{t \leq T}|\theta(t) - \overline{\theta}(x,t)| \geq \eta\} \leq \frac{C(q,Q,T)}{\eta^q}(1 + |y|^{qs\vee 1}) \sum_{k=1}^{m(T)} \gamma_k^{1+q/2}.$$

(We write $P_{x,y}$ *for the law of the Markov chain* (θ_n,y_n) *with initial states* $\theta_0 = x$, $Y_0 = y$.*)*

Remark. Let us observe that if $\gamma_k = \gamma$ constant, the sum $\sum_{k=1}^{m(T)} \gamma_k^{1+q/2} = T\gamma^{q/2}$ is of order $\gamma^{q/2}$.

Proof. We cannot give here all the details of the proof, for which the reader is referred to [20]. There are actually many technical difficulties to overcome. We restrict ourselves to give the main steps. First we can write

$$\theta(t) = \theta_0 + \sum_{k=1}^{m(t)} \gamma_k h(\theta_{k-1}) + \sum_{k=1}^{m(t)} \gamma_k[H(\theta_{k-1}\cdot y_k) - h(\theta_{k-1})]$$

$$+ \sum_{k=1}^{m(t)} \gamma_k^{1+\nu} \tilde{H}(k,\theta_{k-1},Y_k) \tag{3.3.1}$$

and therefore

$$\theta(t) = \theta_0 + \int_0^t h(\theta(s))ds + R(t) \tag{3.3.2}$$

with

$$\begin{aligned} R(t) = & \sum_{k=1}^{m(t)} \gamma_k (H(\theta_{k-1},Y_k).h(\theta_{k-1})) \\ & + \sum_{k=1}^{m(t)} \gamma_k^{1+\nu} \tilde{H}(k,\theta_{k-1},Y_k) - \int_{A(t)}^t h(\theta(s))ds. \end{aligned} \tag{3.3.3}$$

It is clear that the main step in the proof consists in obtaining majorations for $R(t)$, the inequality in the theorem following then from the Gronwall inequality applied to

$$E(|\theta(t) - \overline{\theta}(t,x)|^q) \leq 2^q[\int_0^t K(Q)t|\theta(s) - \overline{\theta}(t,s)|^q ds] + 2^q E(\sup_{t\leq T}|R_t|^q).$$

A key lemma for the proof is the following

<u>Lemma 1</u>. *Let* Q' *be a compact such that* $\beta := \inf\{|x-x'| : x\in Q,\ x' \notin Q'\} > 0$. *Call* $t_{Q'} = \inf\{t_k : \theta_k \notin Q'\}$. *Then*

$$E_{x,y}\{\sup_{s\leq t} R(s)|^q 1_{\{s\leq t_{Q'}\}}\} \leq C(q,Q')(1+t^{q-1})(1+|y|^{qs}) \sum_{k=1}^{m(t)-1} \gamma_k^{(1+q/2)}$$

for all $x \in Q'$ *and some constant* $C(q,Q')$.

In the proof of this lemma it is relatively easy to obtain the required majoration for the second and third terms of the sum (3.3.3), once some majoration has been established for the moments $E(|Y_k|^q)$. Obtaining such a majoration for the moments and dealing with the first sum in (3.3.3) are the big technical difficulties in the proof of the theorem. For the majoration of the moments we refer to [18] Proposition 2.1 or [20], indicating only that for any compact Q one can find a stopping time $\sigma(Q) \leq t_Q := \inf\{n : \theta_n \notin Q\}$ with the properties

$$\sup_n E_{x,y}\{|Y_n|^q 1_{\{n\leq\sigma(Q)\}}\} \leq C(q,Q)(1 + |y|^q), \tag{3.3.4}$$

and for every $u > 1$ with $\rho u \leq p$

$$P_{x,y}\{\sigma(Q) \leq m(T),\ \sigma(Q) < \tau(Q)\} \leq \tilde{C}(Q,u)(1+|y|^{\rho u})(\sum_{n\leq m(T)} \gamma_n^u). \tag{3.3.5}$$

Then one has the following majorations for the first sums in (3.3.3).

Let

$$S^*(t) := \sup_{n\leq m(t)} \left| \sum_{k=1}^{n} \gamma_k[H(\theta_{k-1},Y_k) - h(\theta_{k-1})]1_{\{k\leq\sigma(Q)\}}\right|,$$

then

$$E_{x,y}(|S^*(t)|^q) \leq C(q,Q)(1 + t^{q-1})(1 + |y|^{qs}) \sum_{k=1}^{m(t)-1} \gamma_i^{1+q/2}. \tag{3.3.6}$$

To show (3.3.6) one decomposes the sums in the following way:

$$\sum_{k=1}^{n} \gamma_k[H(\theta_{k-1},Y_k) - h(\theta_{k-1})]1_{\{k\leq\sigma(Q)\}} = M_n + A_n$$

where, using the hypothesis [H.3]:

$$M_n := \sum_{0\leq k\leq n-1} \gamma_k(v_{\theta_k}(Y_{k+1}) - \Pi_{\theta_k}v_{\theta_k}(Y_k))1_{\{k\leq\sigma(Q)\}} \tag{3.3.7}$$

$$A_n := \sum_{0\leq k\leq n-1} \gamma_k(\Pi_{\theta_k}v_{\theta_k}(Y_k) - \Pi_{\theta_k}v_{\theta_k}(Y_{k+1}))1_{\{k\leq\sigma(Q)\}}.$$

One observes that (M_n) is a martingale and the Burkolder-Davis-Gundy inequality gives

$$E(\sup_{0\leq k\leq m(t)} |M_k|^q) \leq C_q E\{(\sum_{k=0}^{m(t)} \gamma_k^2|v_{\theta_k}(Y_{k+1})-\Pi_{\theta_k}v_{\theta_k}(Y_k)|^2 1_{\{k\leq\sigma(Q)\}})^{q/2}\}.$$

Using then the following transform for the latter sums:

$$(\sum_{k=0}^{r} \gamma_k^2 m_k^2)^{q/2} \leq (\sum_{k=0}^{r} \gamma_k)^{(q-2)/2}(\sum_{k=0}^{r} \gamma_k^{1+q/2}m_k^q)$$

and using (3.3.4) one obtains the desired inequality for the sums $E(\sup_{k\leq m(t)} |M_k|^q)$.

To treat the case of the sums A_n one writes

$$A_n = \gamma_0\Pi_{\theta_0}v_{\theta_0}(Y_0) - \gamma_{n-1}\Pi_{\theta_{n-1}}v_{\theta_{n-1}}(Y_n)1_{\{n-1\leq\sigma(Q)\}} + A_n^1 + A_n^2$$

$$A_n^1 := \sum_{k=1}^{n-1} \gamma_k(\Pi_{\theta_k}v_{\theta_k}(Y_k) - \Pi_{\theta_{k-1}}v_{\theta_{k-1}}(Y_k))1_{\{k\leq\sigma(Q)\}}$$

$$A_n^2 := \sum_{k=1}^{n-1} (\gamma_{k+1} 1_{\{k+1\leq\sigma(Q)\}} - \gamma_k 1_{\{k\leq\sigma(Q)\}})\Pi_{\theta_k}v_{\theta_k}(Y_k)$$

and one uses property (H.3 iii) for the sums A_n^1 and the decreasing monotony of (γ_k) for the sums A_n^2.

The Lemma 1 being thus obtained, the Gronwall lemma as explained leads to an estimate:

$$E_{x,y} \sup_{t\leq T\wedge t_Q,} (|\theta(t)-\bar{\theta}(x,t)|^q) \leq \tilde{C}(q,Q,T)(1+|y|^{qs\vee 1})(1+t^{q-1}) \sum_{k=1}^{m(T)-1} \gamma_k^{1+q/2}$$

from where it can be derived

$$P_{x,y}\{t_{Q'} < T\} \le \frac{\tilde{C}(q,Q',T)}{\beta^q}(1+|y|^{qsv1})(1+t^{q-1})\sum_{k=1}^{m(t)-1} \gamma_k^{1+q/2}$$

and then the theorem.

The reader who would like to be supplied with all the details of the proof is referred to [18] or [20].

IV. Asymptotic behaviour of the algorithm [A].

At this point we have to introduce "stability-hypotheses" analogous to the hypothesis (1.3.4) of Theorem 1 (see Remark 1.4.B).

4.1. Stability conditions on the O.D.E.

We assume that θ_* is an asymptotically stable point of the O.D.E. (3.2.1) with domain of attraction D. This means that for any initial point $x \in D$ $\lim_{t\to\infty} \overline{\theta}(x,t) = \theta_*$. More generally we may replace θ_* by a closed set $\Lambda^* \subset D$, the assumption being then that for every $x \in D$ $\lim_{t\to\infty} \sup_{x'\in\Lambda^*} |\overline{\theta}(x,t)-x'| = 0$. (We shall write $\lim_{t\to\infty} d(\overline{\theta}(x,t),\Lambda^*) = 0$).

Another way of expressing this condition is to assume the existence of a function U ("Lyapunov function") with the following properties.

[L.1] U is twice continuously differentiable in D with $U(\theta) = 0$ for every $\theta \in \Lambda^*$ and $U(\theta) > 0$ for all $\theta \in D-\Lambda^*$.

[L.2] $U'(\theta).h(\theta) < 0$ for $\theta \in D-\Lambda^*$.

[L.3] $\lim_{x\to\partial D} U(x) = +\infty$ and $\lim_{\substack{x\in U\\ |x|\to\infty}} U(x) = +\infty$.

4.2. Convergence sets. Use of the Lyapunov function.

For $\beta > 0$ we note $K(\beta) := \{x : U(x) \le \beta\}$. Also $m(n,T) := \max\{k : \gamma_{n+1} + \cdots + \gamma_k \le T\}$. We have then the following basic proposition.

Proposition 1. *Let* U *be the Lyapunov-function associated with* (D,Λ^*) *as in* [L.1], [L.2], [L.3]. *We call* $\tilde{U}_\beta$ *a* C^1*-function on* $\mathbb{R}^d$ *which coincides with* U *on* K_β *such that* $\inf_{x\notin K} \tilde{U}(x) \ge \beta$. *We set*

$$R(\tilde{U},n,T) := \sup_{n\le k\le m(n,T)} \left| \sum_{i=n+1}^{k} (U(\theta_i)-U(\theta_{i-1})-\gamma_i U'(\theta_{i-1}).h(\theta_{i-1})) \right|.$$

If

$$\tilde{\Omega} := \{\lim_{n\to\infty} R(\tilde{U},n,T) = 0, \ \forall T > 0\} \cap \liminf_n \{\theta_n \in Q\},$$

then

$$\tilde{\Omega} \subset \{\lim_{n\to\infty} d(\theta_n,\Lambda^*) = 0\}.$$

Proof. The proof will follow from Lemmas 2 and 3 below:

Lemma 2. *For every* $\delta > 0$ $\tilde{\Omega} \subset \limsup_n \{\theta_n \in K(\delta)\}$.

Proof of Lemma 2. We may clearly assume $\delta < \beta$ and set

$$A = \tilde{\Omega} \cap \liminf_n \{\theta_n \notin K(\delta)\}. \tag{4.1.1}$$

Let

$$r := \sup_n \{\theta_n \in K(\delta)\} \vee \sup_n \{n : \theta_n \notin Q\}$$

$$-\eta := \sup\{h(x).U'(x) : x \in Q - K(\delta)\}.$$

We choose T such that $\eta(T - \gamma_r) \geq 2\beta$. On the set A one has $r < \infty$ and, for all $n > r$

$$R(\tilde{U},n,T) \geq |U(\theta_{n+m(n,T)}) - U(\theta_n) - \sum_{i=n+1}^{m(n,T)} \gamma_i U^1(\theta_{i-1}).h(\theta_{i-1})|$$

$$\geq \beta.\delta > 0.$$

The property $\lim_{n\to\infty} R(\tilde{U},n,T) = 0$ implies $A = \phi$. This gives the lemma.

Lemma 2. *If* $0 < \delta_1 < \delta_2 < \beta$ *then*

$$\tilde{\Omega} \cap \limsup_n \{\theta_n \in K(\delta_1)\} \subset \liminf_n \{\theta_n \in K(\delta_2)\}.$$

Proof of Lemma 2. We set

$$A := \tilde{\Omega} \cap \limsup_n \{\theta_n \in K(\delta_1)\} \cap \limsup_n \{\theta_n \notin K(\delta_2)\}$$

and prove $A = \phi$.

Let

$$-\eta := \sup\{h(x).U'(x) : x \in K(\delta_2) - K(\delta_1)\}.$$

For every $r > \sup\{n : \theta_n \notin Q\}$ we define

$$\theta_r := \inf\{n : n \geq r, \ \theta_n \in K(\delta_1)\}$$

$$\bar{\sigma}_r := \inf\{n : n > \sigma_r \quad \theta_n \notin K(\delta_2)\}$$

$$\lambda_r := \max\{n : \sigma_r \leqq n \leqq \bar{\sigma}_r \quad \theta_n \in K(\delta_1)\}$$

$$\mu_r := m(\lambda_r, T) \quad \text{with} \quad T \quad \text{such that} \quad \eta(T-\gamma_0) > \delta_2 - \delta_1.$$

On A, as a consequence of the definitions: $\sigma_r \leqq \lambda_r \leqq \mu_r \wedge \bar{\sigma}_r < \infty$, and

$$R(\tilde{U}, \lambda_r, T) \geqq |U(\theta_{\mu_r \wedge \bar{\sigma}_r}) - U(\theta_{\lambda_r}) - \sum_{i=\lambda_r+1}^{\mu_r \wedge \bar{\sigma}_r} \gamma_i U'(\theta_{i-1}).h(\theta_{i-1})|$$

$$\geqq \delta_2 - \delta_1. \qquad (4.1.2)$$

In fact, if $\bar{\sigma}_r \leqq \mu_r$, $U(\theta_{\mu_r \wedge \bar{\sigma}_r}) - U(\theta_{\lambda_r}) \geqq \delta_2 - \delta_1$ and, if $\mu_r < \bar{\sigma}_r$, $U(\theta_{\mu_r \wedge \bar{\sigma}_r}) - U(\theta_{\lambda_r}) \geqq 0$ while

$$\sum_{i=\lambda_r+1}^{r} \gamma_i U'(\theta_{i-1}).h(\theta_{i-1}) \geqq \eta(T - \gamma_{\mu_r}) \geqq \delta_2 - \delta_1.$$

The inequality (4.1.2) and the property $\lim_{n\to\infty} R(\tilde{U}, n, T) = 0$ on $\tilde{\Omega}$ imply $A = \phi$. This proves Lemma 3.

The proposition follows clearly from Lemma 2 and Lemma 3 which together imply $\tilde{\Omega} \subset \liminf_{n\to\infty}\{\theta_n \in K(\delta)\}$ for all $\delta > 0$.

4.3. Some asymptotic results.

Theorem 3. *Let us add to the hypotheses of Sections 3.1 and 3.2 the existence of a Lyapunov function as in Section 4.1, let us assume that* $\sum_i \gamma_i^\alpha < \infty$ *for some* $\alpha > 1$ *and also that:*

[B.L] *For every compact set* $S \subset \mathbb{R}^d$ $x \in S$ *and* $y \in \mathbb{R}^K$, $q \leqq p$ *all* n*:*

$$E_{x,y}\{|Y_n|^q \, 1_{\{n \leqq \tau(S)\}}\} \leqq C(S)(1 + |y|^q)$$

where

$$\tau(S) := \inf\{k : \theta_k \in S\}.$$

Then, if Q *is any compact in* D *and* S *any compact with* $S \subset Q \subset \mathbb{R}^d$ *and* $x \in Q$, $P_{x,y}$ *a.s.:*

$$\tilde{\Omega} = \{\tau(S) = \infty\} \cap \limsup_n\{\theta_n \in Q\} \subset \{\lim_{n\to\infty} d(\theta_n, \Lambda^*) = 0\}.$$

Proof. A basic tool of the proof is a majoration lemma analogous to inequality 3.3.6. We state it in the following form:

Lemma 4. *φ being a function with bounded first and second deviations let us set:*

$$\varepsilon_i(\varphi) = \varphi(\theta_i) - \varphi(\theta_{i-1}) - \gamma_i \varphi'(\theta_{i-1}).h(\theta_{i-1})$$

$$S_S^*(\varphi,n,T) = \sup_{n \leq k \leq m(n,T)} \sum_{i=n+1}^{k} \varepsilon_i(\varphi) 1_{\{i \leq \tau(S)\}}$$

Let us assume $p \geq 2\tilde{s}$ with $\tilde{s} := \max(a, b+\rho, 2p)$ and $1 + \frac{p}{2\tilde{s}} \geq \alpha$. Then

$$\lim_{n \to \infty} S_S^*(\varphi,n,T) = 0 \qquad P_{x,y} \text{ a.s.}$$

The proof of the lemma consists in establishing first the following inequality for $2 \leq q \leq \frac{p}{\tilde{s}}$:

$$E_{x,y}|S_S^*(\varphi,n,T)|^q \leq \tilde{C}(S,q,T)(1 + |y|^{q\tilde{s}})(\sum_{i=n+1}^{m(n,T)} \gamma_i^{1+q/2}). \qquad (4.3.1)$$

The reader is referred to [18] Proposition 3.1 for the proof of (4.3.1) which is based on a decomposition of the sums

$$\sum_{i=n+1}^{k} \gamma_i \varphi'(\theta_{i-1}).[f(i-1,\theta_{i-1},Y_i) - h(\theta_{i-1})] 1_{\{1 \leq \tau(S)\}},$$

analogous to the one used in (3.3.3) and (3.3.7) and uses the same type of inequalities (Burkolder-Davis-Gundy inequality etc.). From these on the limit of the lemma is easily derived if one remarks that, if one sets $n_0 := n_1, \dots, n_r := m(n_{r-1},T) \dots$,

$$\sum_{r \geq 0} E_{x,y}(|S^*(n_r,T)|^q) \leq \tilde{C}(Q,q,T)(1+|y|^{q\tilde{s}})(\sum_{i \geq n+1} \gamma_i^{\alpha}) < \infty$$

and therefore $\lim_{n \to \infty} S^*(n_r,T) = 0$ a.s.

To prove the theorem one considers $K_\beta \supset Q$, $\beta' > \beta$ and take $\tilde{U} = U$ on K_β with $\inf_{x \in K_{\beta'}} |\tilde{U}(x)| \geq \beta'$ $\tilde{U}$ being a C^2-function on $\mathbb{R}^d$.

Lemma 4 implies that

$$\{\tau(S) = \infty\} \subset \{\lim_{n \to \infty} R(\tilde{U},n,T) = \lim_{n \to \infty} S_S^*(\tilde{U},n,T) = 0, \forall T > 0\}.$$

In view of Proposition 1 one has only to prove $\tilde{\Omega} \subset \liminf_n \{\theta_n \in K_{\beta'}\}$, or:

$$A := \tilde{\Omega} \cap \limsup_n \{\theta_n \notin K_{\beta'}\} = \phi.$$

One defines $-\eta := \sup\{h(x).U'(x) : x \in K_{\beta'} - Q\}$, takes T such that

$(T - \gamma_0)\eta > \beta' - \beta$ and for every $r \geq 0$ we set

$$\lambda_r := \inf\{n : n \geq r, \ \theta_n \in Q\}$$
$$\sigma_\lambda := \inf\{n : n \geq \lambda_r, \ \theta_n \notin K_{\beta'}\}$$
$$\tilde{\lambda}_r := \max\{n : \lambda_r \leq n \leq \sigma_\lambda, \ \theta_n \in Q\}$$
$$\mu_r := m(\tilde{\lambda}_r, T).$$

Then on A:

$$S_S^*(\tilde{U}, \tilde{\lambda}_r, T) \geq |\tilde{U}(\theta_{\sigma_\lambda \wedge \mu_r}) - \tilde{U}(\theta_{\tilde{\lambda}_r}) - \sum_{i=\tilde{\lambda}_r+1}^{\sigma_r \wedge \mu_r} \gamma_i h(\theta_{i-1}).U'(\theta_{i-1})| \geq \beta' - \beta > 0.$$

But since also $\lim\limits_{\lambda\to\infty} S_S^*(\tilde{U}, \tilde{\lambda}_r, T) = 0$ on A one should have $A = \phi$. This proves the theorem.

Corollary. *If* $D = \mathbb{R}^d$ *and if* $(\theta_n)_{n>0}$ *is bounded a.s. then* $\lim\limits_{n\to\infty} d(\theta_n, \Lambda^*) = 0$, *a.s.*

The case of the equalizer of Example 2.1.A for which $U(\theta)$ is the quadratic form $E|\theta.U - a_n|^2$ with unique minimum θ_* enters in the hypotheses of this corollary. One has only to show the boundedness of (θ_n). (See [5] or [17].)

4.3. Other results.

The proof of Theorem 3 was given under [B.L]. If one drops this hypothesis one has only the weaker moment-inequality given by (3.3.4). In this case one can only state (see [18] Theorem 1).

Theorem 4. *If one drops the assumption* B.L. *of Theorem* 4, *then for each compact* $Q \subset D$ *there exists a constant* C *with the property: for each* N, $x \in Q$, $y \in \mathbb{R}^K$

$$P_{x,y}^N\{\lim_{n\to\infty} d(\theta_n, \Lambda^*) = 0\} \geq 1 - C(1+|y|^{2\tilde{s}(\alpha-1)}) \sum_{i>N} \gamma_i^\alpha,$$

where $P_{x,y}^N$ *denotes the conditional law of* $(\theta_{N+n}, Y_{N+n})_{n\geq 0}$ *given* $\theta_N = x$, $Y_N = y$.

This theorem is a generalization to our setting of a theorem of L. Ljung ([15]).

The corollary above shows the importance of being able to prove the boundedness of $(\theta_n)_{n\geq 0}$. Some conditions can be found in [18] Section 6.

V. Final remarks and conclusions.

We tried in this lecture to make clearly the distinction between two aspects of the study: the study in finite horizon with the help of some averaging principle and the study of the asymptotic behaviour. In the second type of study the general results one can get are somehow quite poor as soon as the (O.D.E.) has several domains of attraction. There is certainly a need for more precise results with more stringent hypotheses motivated by practical situations.

One should also consider that theoretically non convergent algorithms may behave in the following way: for a long time they stay in the vicinity of an attractor of the (O.D.E.) (as if they were to converge to it) before moving to another domain of attraction. Such algorithms "Practically converge to an attractor". This underlines the interest of the study in finite horizon. We did not mention here in this lecture the gaussian approximation theorems of the literature (see [2] or [11]). We should also emphasize that at the present time there seem to be *very few* large deviation theorems connected with the problem we just mentioned. We give the reference [2] which concerns a Robbins-Monro algorithm and there is a brief chapter in [11] devoted to the subject. This is certainly a direction for further researches of mathematical and practical interest.

BIBLIOGRAPHY

[1] A. Benveniste, M. Goursat, G. Ruget, Analysis of stochastic approximation schemes with discontinuous dependent forcing terms. I.E.E.E. Trans. Autom. Control, Vol.AC. 25 n° 6 (1980).

[2] C. Bouton, Approximation gaussienne d'Algorithmes sothastiques. These 3 Cycle, Paris VI (1985).

[3] M. Cottrell, J.C. Fort, G. Malgouyres, Large deviations and rare events in the study of stochastic algorithms. I.E.E.E. Trans. Autom. Control, Vol. AC. 28 n° 3 (1983), pp.907-920.

[4] O.P. Deveritskii, A.L. Fradkov, Two models for analysing the dynamics of adaptation algorithms. Automatika. I. Telemekhanika, n° 1 (1974), pp.66-75.

[5] E. Eweda, O. Macchi, Convergence analysis of an adaptive linear estimation algorithm. I.E.E.E. Trans. Autom. Control, Vol. AC. 29 n° 2 (1984), pp.119-123.

[6] E. Eweda, O. Macchi, Convergence analysis of self adaptive equalizers. I.E.E.E. Trans. Inf. Theory, Vol. IT-30 n° 2 (1984), pp.161-176.

[7] V. Fabian, Stochastic approximation. In optimizing methods in statistics (Editor: J.S. Rustagi). Acad. Press, 1971, pp.439-470.

[8] E.G. Gladyshev, On stochastic approximation. Theory of Probability and its applications. Vol. 10 (1985), pp.275-278.

[9] C.C. Heyde, D. Hall, Martingale limit theory and its applications. Acad. Press, 1980.

[10] J. Kiefer, J. Wolfowitz, Stochastic estimation of the modulus of a regression function. Ann. Math. Stat. Vol. 23 (1952), pp.462-466.
[11] H. Kushner, Approximation and weak convergence methods for random processes, M.I.T. Press. Cambridge, 1984.
[12] H. Kushner, D.S. Clark, Stochastic approximation methods for constrained and unconstrained systems.
[13] H. Kushner, Stochastic approximation with discontinuous dynamics and state dependent noise. J. of Math. An. and Appl. Vol. 82 (1981), pp.527-542.
[14] H. Kushner, A. Shwartz, An invariant measure approach to the convergence of stochastic approximations with state dependent noise. S.I.A.M. 5. Control. Opt. Vol. 22 n° 1 (1984).
[15] L. Ljung, Analysis of recursive stochastic algorithms. I.E.E.E. Trans. Autom. Control. Vol. AC. 22 n° 2 (1977), pp.551-575.
[16] L. Ljung, T. Söderström, Theory and practice of recursive identification. M.I.T. Press, 1983.
[17] M. Métivier, P. Priouret, Application of a Kushner and Clark. -- Lemma to general classes of stochastic algorithms. I.E.E.E. Trans. Inf. Theory. IT-30 (1984), pp.140-150.
[18] M. Métivier, P. Priouret, Théorèmes de convergence presque sure pour une classe d'algorithmes stochastiques. Preprint, Ecole Polytechnique, Palaiseau, (1984).
[19] M. Métivier, P. Priouret, Convergence avec probabilité (1-ε) d'algorithmes stochastiques. Application au cas de l'egaliseur aveugle. To appear in Proceedings of the "Seminaire Quantification et Adaptativité". Nice Mai 85, Annales des Télécommunications.
[20] M. Métivier, Introduction to stochastic approximation and adaptive filtering. Lecture Notes. Dipartimento di Matematica, Universita di Rome.
[21] M. Métivier, Semimartingales. De Gruyter-Berlin, 1983.
[22] M.B. Nevelson, R.Z. Has'Minski, Stochastic approximation and recursive estimation. A.M.S. Translations, Vol. 42.
[23] H. Robbins, S. Monro, A stochastic approximation method. Ann. Math. Stat. Vol. 22 (1951), pp.400-407.
[24] D. Ruppert, Almost sure approximation to the Robbins-Monro and Kiefer-Wolfowitz processes with dependent noise. The Ann. of Prob. Vol. 10 n° 1 (1982), pp.178-187.
[25] D. Ruppert, Convergence of stochastic approximation algorithms with non additive dependent disturbances and applications. Preprint (1983).
[26] M.T. Wasan, Stochastic approximation. Cambridge University Press (1968).
[27] M.I. Freidlin, Random perturbations of dynamical systems. Springer, 1984.

ESTIMATION THEORY AND STATISTICAL PHYSICS

S.K. Mitter
Department of Electrical Engineering and Computer Science and
Laboratory for Information and Decision Systems, M.I.T.
Cambridge, MA 02139

1. INTRODUCTION

In my previous work on non-linear filtering for diffusion processes [Mitter 1980,1983] I have discussed the analogies that exist between these problems and problems of quantum physics from the Feynman point of view [cf. Glimm-Jaffe 1981]. The basic idea here is that construction of a non-linear filter involves an integration over function space which is exactly analogous to the construction of a measure on path-space via the Feynman-Kac-Nelson Formula. According to this viewpoint, the Kalman-Bucy filtering problem, namely the filtering of Gauss-Markov processes in the presence of additive white Gaussian noise occupies the same role as the Ornstein-Uhlenbeck process (finite or infinite-dimensional) in Quantum Mechanics or Quantum Field Theory. That this analogy may be deeper is borne out by the fact that a solvable Lie algebra, the oscillator algebra which contains the Heisenberg algebra as a derived algebra is intrinsically attached to the Kalman-Bucy Filtering problem. I have also shown that the problem of non-linear filtering of diffusion processes admits a stochastic variational interpretation [Fleming-Mitter 1982]. The objective of this paper is to strengthen these analogies further with a view to showing the close relationship of estimation theory to statistical mechanics. The motivation for this comes from problems of estimation and inverse problems related to image processing.

In order to carry out this program it is necessary to generalize these ideas to filtering problems for infinite-dimensional processes where we are forced to work in the context of generalized random fields. There are two types of processes involved: continuous processes such as infinite-dimensional Ornstein-Uhlenbeck processes and their L^2-functionals which represent intensities of images and processes of a "discrete"-nature which will represent "boundaries" of images. The most interesting models are obtained when these processes are coupled according to a probabilistic mechanisms. The "discrete" processes should be thought of as gauge fields and will be a process on connection forms.

Although the estimation problems of interest are naturally viewed in the context of random fields which are independent of time, the best way to obtain sample functions for these fields is to simulate it via finite or infinite-dimensional stochastic differential equations whose invariant distribution coincides with the distribution of the time-independent random field. This Monte-Carlo simulation procedure is the same idea as stochastic quantization, an idea advanced by Parisi (cf. Parisi-Wu 1981) and recently studied in a rigorous manner for $(P\phi)_2$ fields in a finite volume by Jona-Lasinio and Mitter, 1984. The problem which are of interest here are filtering problems associated with these stochastic fields obtained by introducing observations which are "local" and studying the behaviour of these filters as $t \to \infty$. To make any progress however, one would have to work with lattice approximations of these stochastic fields and reduce the filtering problems to a finite dimensional situation and even here there are severe technical difficulties. When the observations are however local and considered to be on the stationary Gibbs field the problems amounts to looking at the invariant distribution of a stochastic differential equation with a coupling coming from the observations.

The main objective of this paper is to explore these relationships between problems of estimation and stochastic quantization (stochastic mechanics) at a conceptual level. The detailed technical discussion will appear elsewhere.

2. SIGNAL MODELS FOR IMAGE PROCESSING

In order to treat problems of Image Processing in a probabilistic framework we need probabilistic models for the signals in question. These signals are various attributes of images such as intensity of images and boundaries between smooth parts of images. The probabilistic models we choose are Gibbsian random fields and are borrowed from statistical mechanics. These models for image processing have been used by several authors recently (c.f. Geman and Geman 1984, Grenander 1984, Marroquin 1985). The exposition of signals models given below follows Sinai 1982. The signal models we wish to consider correspond to statistical mechanical models on a finite lattice and we shall take this lattice to be Z^2 with the Euclidean norm. The sample space Ω consists of functions $\phi: Z^2 \to \Phi: x=(x_1,x_2) \to \phi(x)$, where Φ is a finite set, a homogeneous space of a compact Lie Group with the natural σ-algebra of Borel Sets or R^1. The sample space Ω is termed a configuration space in statistical physics. For $V \subset Z^2$, a finite subset, we denote by $\phi(V) = \{\phi(x) | x \varepsilon V\}$ and $V \; Z^2 = \{\phi(V) | V \subset Z^2\}$, finite.

For a non-empty finite subset $V \subset Z^2$ we are given a function $I:\Omega(V) \to R:\phi(V) \to I(\phi(V))$ which is called the potential. $I(\phi(V)$ is the joint interation energy of $\phi(x)$ inside the domain V.

For an arbitrary finite $W \subset Z^2$, we define the energy $H(\phi(w))$ of the configuration ϕ in the domain W as

$$H(\phi(w)) = \sum_{V\ W} I(\phi(v)). \tag{2.1}$$

The sum

$$H(\phi(w)\,|\,\phi(Z^2\backslash W)) = \sum_{\substack{V\cap W=\phi \\ V\cap Z^2\backslash W=\phi}} I(\phi(V)) \tag{2.2}$$

is the interaction energy between the configuration $\phi(W)$ and $\phi(Z^2\backslash W)$, where $\phi(Z^2\backslash W)$ is the boundary condition. The total energy of the configuration $\phi(W)$ is the sum $H(\phi(W)) + H(\phi(Z^2\backslash W))$.

The Hamiltonian is defined as the formal series.

$$H(\phi) = \sum_{V} I(\phi(V)), \tag{2.3}$$

where V ranges over all finite subsets of Z^2. The model of most interest to us is the 2-dimensional Ising model when $\Phi = \{1, -1\}$ (spins), and $I(\phi(V)) = 0$ unless $V = \{x,y\}$ with $||x-y|| = 1$.

We take $I(\phi(V)) = J\phi(x)\phi(y)$ where $J = \pm 1$.

$$H(\phi) = -J \sum_{\substack{\{x,y\} \\ ||x-y||=1}} \phi(x)\phi(y) \tag{2.4}$$

This Hamiltonian is translation invariant and reflection invariant. If J=+1, the model is ferromagnetic and J=-1 corresponds to an antiferromagnetic model.

The other model which will be of interest to us is the Ising model with an external magnetic field with Hamiltonian

$$H(\phi) = -J \sum_{\substack{\{x,y\} \\ ||x-y||=1}} \phi(x)\phi(y) - h \sum_{x\varepsilon Z^2} \phi(x). \tag{2.5}$$

where h may be random.

Finally a case of interest to us is the Hamiltonian

$$H(\phi) = \sum_{\substack{\{x,y\} \\ ||x-y||=1}} J(x,y)\phi(x)\phi(y) - h \sum_{x \in Z^2} \phi(x) \tag{2.6}$$

where J(x,y) is random. This corresponds to a spin glass.

Let a measure be given on Φ and for every $V \subset Z^2$ we consider the product measure

$$\prod_{s \in V} d\chi(\phi(s)) = d\mu.$$

We are interested in Gibbs distributions in the domain V which is a probability distribution on $\Omega(V)$ whose density with respect to $d\mu$ is given by

$$\frac{\exp(-H(\phi(V)))}{\int \exp(-H(\phi(V))d\mu} \tag{2.7}$$

$Z = \int \exp(-H(\phi(v))d\mu$ is called the partition function.

This corresponds to the so-called Gibbs distributions in V with free boundary conditions (Dirichlet Boundary conditions in the literature of quantum field theory). The signals ϕ we shall be interested in will have Gibbs distributions given by (2.7) and is defined by prescribing the potential I.

For modelling intensities of images we shall typically use a Gibbs distribution corresponding to a Hamiltonian of the 2-dimensional Ising model. Boundaries between smooth patches of images will be modelled as Gibbs distributions on the dual lattice. We shall discuss this in a later section.

2.1 Simulation of the Gibbs Distribution

Sample functions of the Gibbs distribution are obtained by constructing a Markov chain whose states correspond to the configurations of the Lattice field at time points 1,2,... in such a way that it has a unique invariant measure as the Gibbs measure $\exp(-H(\phi(V)))d\mu$. This chain clearly has to be reversible. Various algorithms for creating such a chain are known of which the Metropolis algorithm is the best known. In practice one may have to deal with a very large subset of a lattice and the random variables at the lattice points may take values in R^1. In this case it may be useful to make a diffusion approximation and simulate a stochastic differential equation with the same properties as above. This is the analogy to stochastic quantization we referred to before.

One therefore has to study the stochastic differential equation of the basic form:

$$d\phi(t,x) = -D_\phi H(\phi(t,x)) + dw(t,x)$$
$$\phi(0,x) = \phi_0(x) \tag{2.8}$$

where in general $\phi(t,\cdot)$ is a generalized random field and D_ϕ denotes "functional" derivative with respect to ϕ. The questions of interest is to construct the measure in the path space of ϕ and prove that the stochastic differential equation (2.8) has a unique invariant measure with density (with respect to an appropriate measure) $\exp(-H(\phi))$. The interest in this model for generating Gibbs distributions is that only Gaussian random numbers need to be generated and the computation is amenable to parallel processing.

3. STOCHASTIC MECHANICS, STOCHASTIC QUANTIZATION AND SIMULATION OF IMAGE INTENSITIES

We start with some well-known facts relating the Feynman-Kac Formula and the Girsanov Formula. (Carmona 1979, Simon 1980, Mitter 1980).

Let us suppose that $V:R^n \to R$, be measurable, bounded below and tends to $+\infty$ as $|x| \to \infty$ and consider the Schrödinger operator $H = -\Delta + V$ where Δ is the n-dimensional Laplacian. Then H defines a self-adjoint operator on $L^2(R^n; dx)$ which is bounded below and the lower bound λ (assumed to be 0) of the spectrum of H is an eigenvalue of H. Let $\psi(x)$ be the corresponding eigenfunction of H, the so-called ground state and assume $\psi(x) > 0$. We normalize $\psi(x)$ i.e. $\int|\psi(x)|^2 dx = 1$. Define the probability measure $d\mu = |\psi(x)|^2 dx$, and the unitary operator

$$U : L^2(R^n; dx) \to L^2(R^n; d\mu(x))$$
$$: f \to \psi^{-1} f.$$

If we define the Dirichlet form for $f, g \in C_c^\infty(R^n)$

$$\delta(f,g) = \frac{1}{2}\int_{R^n} \nabla f(x) \cdot \nabla g(x) dx$$

then a calculation shows

$$\delta(f,g) = (\mathcal{L}f,g)_\mu$$

where $(\, , \,)_\mu$ denotes the scalar product in $L^2(R^n; d\mu)$ and $\mathcal{L}$ is the diffusion operator (self-adjoint, positive)

$$\mathcal{L}\phi = -\frac{1}{2}\Delta\phi + \nabla b \, . \, \nabla\phi \quad , \; b := -\log\psi \tag{3.1}$$

Since φ satisfies the equation

$$\frac{1}{2}\Delta\varphi + V(x)\varphi = 0, \tag{3.2}$$

in the sense of distributions (note that we have taken $\lambda=0$), a direct calculation shows

$$V(x) = \frac{1}{2}(|\nabla b(x)| - \Delta b(x)). \tag{3.3}$$

Now using the Feynman-Kac formula for (3.2)

$$\varphi(x) = E^{\omega}[\varphi(x(t))\exp(-\int_0^t V(x(s))ds)|x(t) = x], \tag{3.4}$$

where E^{ω} denotes expectation w.r. to Wiener measure, the properties of V, equation (3.3) and the generalized Ito-differential rule (Meyer 1978), we see

$$\begin{aligned} L(t) &= \varphi^{-1}(x(0)\varphi(x(t))\exp(-\int_0^t V(x(x))ds) \\ &= \exp[-\int_0^t \nabla b(x(s)).dx(s) - \frac{1}{2}\int_0^t |\nabla b(x(s))|^2 ds] \end{aligned} \tag{3.5}$$

is a $(\Omega, \mathcal{F}_t, \mu^{\omega})$ martingale, where $\mathcal{F}_t$ is the σ-field generated by $(x(s)|0 \leq s \leq t)$ and μ^{ω} denotes Wiener measure. Therefore the process $(w(t)|t \geq 0)$ defined by

$$w(t) = x(t) - x(0) + \int_0^t \nabla b(x(s))ds \tag{3.6}$$

is standard Brownian motion with respect to the measure $\mu^{\mathfrak{X}}$ defined by

$$\frac{d\mu^{\mathfrak{X}}}{d\mu^{\omega}} = L(t) \tag{3.7}$$

Hence $x(t)$ considered as a stochastic process on $(\Omega, \mathcal{F}_t, \mu^{\mathfrak{X}})$ is a weak solution of the equation

$$\begin{aligned} dx(t) &= -\nabla b(x(t))dt + dw(t) \\ x(0) &= x. \end{aligned} \tag{3.8}$$

Indeed, the stochastic process defined by (3.8) is a Feller process, which is recurrent and has μ as its unique invariant measure. Therefore with the ground state of a Schrodinger operator we have attached a diffusion process which is ergodic.

The converse procedure is of interest to us. Suppose we start with the stochastic differential equation on R^n given by (3.8) where $-\nabla b(.)$ is a singular drift. The case of interest to us is where b(.) is a polynomial which is bounded below. Now, typically the Novikov condition

$$E \exp\left(\frac{1}{2} \int_0^t |\nabla b(x(s))|^2 ds\right) < \infty \tag{3.9}$$

will fail for these drifts and hence the Girsanov functional

$$L(t) = \exp\left(- \int_0^t \nabla b(x(s)).dx(s) - \frac{1}{2} \int_0^t |\nabla b(x(s))|^2 ds\right) \tag{3.10}$$

although a super-martingale, need not satisfy

$$E^{\mu}_{\mathfrak{X}} [L(t)] = 1$$

and hence fail to be a martingale. Therefore, the method of proof to show that equation (3.8) has a weak solution using the Girsanov formula will not work. To show, however that (3.8) has a weak solution and the process defined by (3.8) is recurrent and possesses a unique invariant measure, we consider the operator

$$= - \frac{1}{2} \Delta + \nabla b.\nabla, \tag{3.11}$$

and transform it to

$$H = - \frac{1}{2} \Delta + V(x), \tag{3.12}$$

where

$$V(x) = \frac{1}{2} (|\nabla b(x)|^2 - \Delta b(x)),$$

using the Gauge transformation,

$$\phi(x) \rightarrow \exp(b(x))\phi(x). \tag{3.13}$$

Then using functional-analytic arguments (cf. Segal 1970), one shows that the semigroup e^{-tH} is indeed, a strongly continuous, self-adjoint contractive semigroup on $L^2(d\mu)$ where μ is an appropriate probability measure, $e^{-tH}1=1$, is positivity preserving and improving and 1 is the unique ground state. It then follows that

e^{-tH} is ergodic with μ as its unique invariant measure. To carry out this program rigorously, one would use hyper-contractive estimates of Nelson and Segal and hence it is natural to work with the stochastic differential equation:

$$dx(t) = -\nabla b(x(t))dt + d\xi(t), \tag{3.14}$$

where

$$d\xi(t) = -A\xi(t)dt + dw(t), \tag{3.15}$$

where A is a symmetric nxn matrix. Therefore (3.15) defines a generalized Ornstein-Uhlenbeck process. Let us define formally the semigroup,

$$(e^{-tL}f)(x) = E^{\omega}[f(x(t))L(t)R(t)\,|x(t) = x], \tag{3.16}$$

where

$$L(t) = \exp(-\int_0^t \nabla b(x(x)).dx(s) - \frac{1}{2}\int_0^t |\nabla b(x(s))|^2 ds)$$

$$R(t) = \exp(-\int_0^t Ax(s).ds(s) - \frac{1}{2}\int_0^t |Ax(s)|^2 ds)$$

and E^{ω} denotes expectation with respect to Wiener measure. One wants to write (3.16) in Feynman-Kac form, i.e. in the form

$$(e^{-tL}f)(x) = E^{\omega}[f(x(t))\exp(-\int_0^t V(x(s)ds)\,|x(t) = x] . \tag{3.17}$$

To do this we need an Ito differential rule for b. For this we need that b is continuous,

$$\nabla b \in L^2_{loc.}(R^n; dx) \text{ and } \Delta b \in L^1_{loc.}(R^n; dx).$$

The measure $d\mu$ referred to previously then is

$$d\mu = \exp(-\int_0^t V(x(s))ds)d\mu^{\xi},$$

where

$$\frac{d\mu^{\xi}}{d\mu^{\omega}} = R(t).$$

The rest then follows from Nelson 1973, Segal 1970, Glimm-Jaffe 1981, Gross 1972.

These ideas can be generalized to random fields (cf. Jona Lasinio-Mitter loc.cit). Let Λ R^2 be a square and consider the Laplacian Δ on R^2 with Dirichlet Boundary conditions, which is a self-adjoint operator on $V = L^2(\Lambda)$. Let $H^1(\Lambda)$ be the Sobolev space of functions f on R^2 with norm

$$||f||^2_{H^1} = \int_\Lambda \left|(-\Delta+1)^{1/2} f(x)\right|^2 dx$$

and let $V' = H^{-1}(\Lambda)$ be the dual-space of distributions with norm

$$||\phi||^2_{H^{-1}} = \int_\Lambda |(-\Delta+1)^{-1/2}\phi(x)|^2 dx.$$

Let μ_c denote the Gaussian measure on V' with mean zero and covariance operator C given by

$$C = (-\Delta+1)^{-1}. \tag{3.19}$$

Consider the stochastic differential equation on V':

$$\begin{aligned} d\phi(t) &= -\frac{1}{2} C^{-\varepsilon}\phi(t)dt + dw(t), \\ \phi(0) &= \phi,\ 0 < \varepsilon < \frac{1}{2} \end{aligned} \tag{3.20}$$

interpreted in the weak sense, where w(t) is a V'-valued Brownian motion with covariance $C^{1-\varepsilon}\min(t,s)$ (defined using test functions).

One can show that this stochastic process defines a measure on the path space $C(0,\infty;V')$ and we denote this measure by P. Define the semigroup

$$(e^{-tL_0} f)(\phi) = E[f(\phi(t))\,|\phi(t) = \phi], \tag{3.21}$$

where f is a "suitable" function and E denotes expectation w.r. to P-measure. One shows using the work Segal and Nelson referred to above that:

(i) e^{-tL_0} is a strongly continuous semigroup on $L^2(d\mu_c)$, which is contractive, self-adjoint.

(ii) is positivity preserving and improving on $L^2(d\mu_c)$; $e^{-tL_0}1 = 1$.

(iii) is a contraction on $L^p(d\mu_c)$; $1 \leq p < \infty$

(iv) is hypercontractive: $||e^{-tL_0}\phi||_{L^4(d\mu_c)} \leq C||\phi||_{L^2(d\mu_c)}$

for $\phi \in L^2(d\mu_c)$ and t sufficiently large. Moreover the stochastic process defined by (3.20) is ergodic and has μ_c as its unique invariant measure. Therefore the Ornstein-Uhlenbeck process defined by (3.20) is the stochastic quantization of the free euclidean field. In view of the well known relationship between the free field and the Ising model via the lattice approximation (cf. Guerra-Rosen-Simon 1975), we see that simulating the stochastic differential equation (3.20) is a powerful way of obtaining sample functions of the free field (Ising field) when the lattice is large. This can be accomplished in a parallel machine using multi-grid methods.

As remarked by Guerra-Simon-Rosen, the Ising (nearest-neighbour) nature of the lattice fields are not destroyed if we use non-gaussian random variables at the lattice points. This can be accomplished by studying the stochastic differential equation on V'

$$\begin{aligned} d\hat{\phi}(t) &= -\frac{1}{2} C^{-\varepsilon}\hat{\phi}(t)dt + \lambda C^{1-\varepsilon}:\hat{\phi}(t)^3:dt + dw(t) \\ \hat{\phi}(0) &= \phi,\ 0 < \varepsilon < \frac{1}{10} \end{aligned} \tag{3.22}$$

where : : denotes Wick ordering with respect to the covariance C (cf. Glimm-Jaffe 1981). Using an appropriate Ito rule (proved by approximation using tame functions), considering an aprpoximation of (3.22) using a spectral basis related to $(-\Delta+1)$ on $L^2(\Lambda)$, and using limiting arguments coupled with hyper-contractive estimates (see the discussion in the first part of this section), Mitter-Jona Lasinio show that the semi-group defined by

$$(e^{-tL}f)(\phi) = E(f(\phi(t))\exp(\xi(t))\,|\phi(t) = \phi) \tag{3.23}$$

where

$$\xi(t) = -\frac{\lambda}{2}\int_0^t (:\phi(s)^3:,\ dw(s)) - \frac{\lambda^2}{8}\int_0^t (:\phi(s)^3:,\ C^{1-\varepsilon}:\phi(s)^3:)ds \tag{3.24}$$

where $\phi(t)$ is the Ornstein-Uhlenbeck process defined by (3.20) satisfy

(i) $e^{-tL}\,1=1$ and e^{-tL} is a contraction on $L^\infty(d\mu_c)$.

(ii) e^{-tL} is a strongly continuous, self-adjoint semigroup on $L^2(d\mu)$ with $d\mu = \exp(-\frac{\lambda}{4}\int_\Lambda :\phi^4(x):dx)d\mu_c$, positivity preserving and improving and 1 is the unique ground state.

(iii) e^{-tL} is ergodic and mixing.

This then allows them to show that the stochastic differential equation (3.22) has a weak solution and has as its unique invariant measure μ.

Since the stochastic differential equations under consideration define a Markov process which is ergodic (and mixing), we have

$$\lim_{t\uparrow\infty} E_\phi((\hat{\phi}(t),\ f_1)(\hat{\phi}(t),f_2)\ \ldots\ (\hat{\phi}(t),f_n))$$

$$= \frac{\int(\phi,f_1)\ldots(\hat{\phi},f_n)d\mu}{\int d\mu}$$

where the f_i are test functions. This enables us to compute spatial statistics of the time independent random field from the simulation of the stochastic differential equation.

4. Some Estimation Problems for Random Fields on a Lattice (cf. Marroquin 1985)

Let $V \subset Z^2$ be a finite subset. Consider a random field

$$f:V \to \Phi \tag{4.1}$$

with Gibbs distribution having a density with respect to $d\mu$ (see (2.7)) given by

$$P_f = Z^{-1}\exp(-\frac{1}{T_0}H_0(f)), \tag{4.2}$$

and $T_0 > 0$ is a parameter (temperature).

We observe a corrupted version of f given by

$$g(j) = \Phi[G(j;f), \nu(j)], \quad j \in S \subset V, \tag{4.3}$$

where H(j,.) is a function with "local" support and Φ is invertible in the sense that $\nu(j) = \varphi^{-1}(g(j), G(j;f))$. We shall assume that $\nu(i)$ and $\nu(j)$ are indepenent and also $\nu(.)$ and f(.) are independent. Let us suppose that the distribution of $\nu(.)$ with respect to $d\mu$ has density

$$p_\nu > 0 \tag{4.4}$$

Define the functions

$$K(i;f,g(i)) = -\ln p_\nu(\Phi^{-1}(g(i), G(i;f)). \tag{4.5}$$

Then the conditional density $p_{f/g}$ can be written as

$$p_{f/g} = Z_p^{-1} \exp(-H_p(f,g)), \tag{4.6}$$

$$H_p(f,g) = \frac{1}{T_0} H_0(f) + \sum_{i \in S} K(i;f,g(i)) \tag{4.7}$$

and Z_p is a normalizing constant.

We can now provide a physical interpretation of the posterior distribution, by considering that, while the prior distribution (4.2) describes the behavior of a free field in thermal equilibrium, the distribution (4.7) describes the behavior of the same field coupled with a fixed (but spatially varying) external field whose value is given by g. The functions K whose magnitude depends on the noise variance, can then be interpreted as the coupling strengths between the two fields. This coupled system is also Gibbsian and if

$$G(i;f) = G(i;f(i))$$

the Markovian structure of this field will be identical to that of the original field.

The importance of this interpretation lies in the fact that the optimal estimate of f can be obtained simply by observing the equilibrium behavior of this coupled field.

In particular if $H_0(f)$ is the Hamiltonian of a ferromagnetic Gaussian Ising field and the observation is,

$$g(i) = f(i) + n(i), \tag{4.8}$$

with n(i) Gaussian, then the coupled field corresponds to a Ising field coupled to a random external field.

The estimates that are of interest to us depends on the choice of the problem. The two most important estimates for image reconstruction purposes are:

(i) $\hat{f} = E(f|g)$ (Conditional mean)

(ii) $\hat{f} = \text{Arg Max } p_{f|g}(f;g)$ (MAP)

From (4.6) and (4.7) the MAP estimate corresponds to minimizing the Hamiltonian $H_p(f,g)$ with respect to f.

4.1 Block Spin Transformation for MAP Estimation (Marroquin 1985)

In order to illustrate the analogy with Statistical Physics further we consider the MAP estimation of a binary Ising field withe the observations taken as the output of a binary symmetric channel with error rate ε. Therefore $\Phi = \{1,-1\}$ and the observation model is given by

$$P(g(i)|f(i)) = \begin{cases} 1-\varepsilon & \text{if } g_i = f_i \\ \varepsilon & \text{if } g_i \neq f_i. \end{cases}$$

Then it is easy to see that

$$H_p(f,g) = \frac{1}{T_0} \sum_{\substack{\{i,j\} \\ ||i-j||=1}} f(i)f(j) + \alpha \sum_i f(i)g(i) \tag{4.8}$$

and

$$\alpha = \ln(\frac{1-\varepsilon}{\varepsilon}).$$

Minimizing H_p is now a combinatorial optimization problem.

4.1.1 Simulated Annealing and Global Minimization

Simulated annealing is a new technique, developed by Kirkpatrick et al (1983) for the solution of combinatorial optimization problems. It is based on the idea that any cost functional of N variables, each of which can take values on some finite set, can be considered as the Hamiltonian (Energy) of a physical system whose state corresponds to a particular value of these variables. Therefore, we can use, say, the Metropolis algorithm to generate, at any given "temperature" T (which now becomes a parameter of the optimization process) samples from the corresponding Gibbs measure. As $T \downarrow 0$ this measure should converge to a measure which concentrates on the states of minimum energy, the state of the system in thermal equilibrium at zero temperature will correspond to the value of f that minimizes the energy H(f) globally.

One serious difficulty, however, is that attaining thermal equilibrium might take a very long time at low temperatures. Kirkpatrick's idea was to start at a relatively high temperature (where thermal equilibrium is reached very fast), and then, to slowly cool the system, until "freezing" occurs and the state stops changing.

The analysis of this algorithm is presented in the Appendix.

4.1.2 Block Spin Transformations

In the case of the MAP estimator, the efficiency of the Simulated Annealing algorithm for the minimization of H_p can be improved by defining large "blocks" of sites (in a manner that is reminiscent of the "block-spin" strategy used by Wilson (1975) in connection with the renormalization group approach to the study of critical phenomena); the optimal estimate for the average value of the field in each of these blocks is found, and then progressively refined by subdividing the blocks in successive annealing stages. We will now show that, if we use a maximum entropy assumption, the structure of the MAP estimation process for Ising models is invariant under the "blocking" transformation; this means that the ground state (i.e., the MAP estimator) of the aggregated process (with blocks of size L) also corresponds to that of an Ising model with a coupled external field, in which the natural temperature is scaled by a factor of 1/L, and the noise (coupling) parameter by a factor of L^2. As a consequence of this scaling, the final temperature for the simulated annealing of this smaller network will be approximately L times larger than for the original problem.

If we denote by $V(f(i), f(j)) = f(i)f(j)$ and $q(f(i), g(i)) = f(i)g(i)$, let $V_c(f(i), f(j))$ and $q_c(f(i), g(i))$ be the extension to RxR of V and q respectively. We then write

$$H_p(f,g) = \frac{1}{T_0} \sum_{\substack{i,j=1 \\ ||i-j||=1}} V_c(f(i),f(j)) + \alpha \sum_i q_c(f(i),g(i)) \tag{4.9}$$

We will now derive an expression for the energy in the "block spin" case. Let us partition the original lattice into square blocks of side L. The "block observations" g_L will now be the density of 1's on each block, i.e.,

$$g_L(i) = \frac{1}{L^2} = \sum_{j \varepsilon B_i} g(j) .$$

where B_i is the ith block. The "block field" f_L is defined in a similar way.

For a given f_L, we compute the energy by assuming a maximum entropy configuration, which occurs when the 1's that correspond to the given density $f_L(i)$ are randomly distributed within the block. The energy will have three terms:

1. Interactions between adjacent blocks:
The interaction between two adjacent blocks i and j will be:

$$I_{ij} = [-1 \cdot (P_{11} + P_{00}) + 1 \cdot (P_{10} + P_{01})] \cdot L$$

where P_{kl} is the probability of having an element with state k on block i adjacent to an element with state l on block j:

$$P_{11} = f_1(i) f_L(j)$$

$$P_{01} = f_L(j)(1 - f_L(i))$$

$$P_{10} = f_L(i)(1 - f_L(j))$$

$$P_{00} = (1 - f_L(i))(1 - f_L(j))$$

Substituting these values we get:

$$I_{ij} = L[2(f_L(i) + f_1(j)) - 4f_L(i)f_L(j) - 1]$$

2. Interactions within each block: the internal interaction I_i is:

$$I_i = 2L(L-1)(-4f_L(i)^2 + 4f_L(i) - 1)$$

3. Interaction with the observations:
Assuming that the 1's in the observations and in the field are independently distributed we get:

$$I_{obs}(i) = \alpha L^2[f_L(i)(1 - g_L(i)) + (f - f_L(i))g_L(i)] =$$

$$= \alpha L^2[f_L(i) + g_L(i) - 2f_L(i)g_L(i)]$$

Finally, the Hamiltonian takes the form

$$H_L(f_L) = \frac{1}{T_0} \sum_{i,j} I_{ij} + \sum_i (\frac{1}{T_0} I_i + I_{obs}(i)) =$$

$$= L\{\frac{1}{T_0} \sum_{i,j} [2(f_L(i) + f_L(j)) - 4f_1(i)f_L(j) - 1] +$$

$$+ \frac{2}{T_0} (L - 1) \sum_i (-4f_L(i)^2 + 4f_L(i) - 1) +$$

$$+ \alpha L \sum_i (f_L(i) + g_L(i) - 2f_L(i)g_L(i)\}$$

note that the sums are taken over pairs of adjacent blocks, and over all the blocks, respectively. For $L = 1$, this expression reduces to (4.9) with

$$V_c(f(i),f(j)) = 2(f(i) + f(j)) - 4f(i)f(j) - 1$$

$$q_c(f(i), g(i)) = f(i) + g(i) - 2f(i)g(i).$$

For $L > 1$, the quadratic terms of H_L are:

$$\frac{L}{T_0} [-4 \sum_{i,j} f_L(i)f_L(j) - 8(L - 1) \sum_i f_L(i)^2]$$

and since

$$- 2 \sum_{i,j} f_L(i)f_L(j) + 2 \sum_i f_L(i)^2 = \sum_{ij} (f_L(i) - f_L(j))^2 \geq 0$$

it follows that

$$\sum_i f_L(i)^2 \geq \sum_{i,j} f_L(i)f_L(j)$$

and

$$-4 \sum_{i,j} f_L(i) f_L(j) - 8(L-1) \sum_i f_L(i)^2 <$$

$$< -(4 + 8(L-1)) \sum_i f_L(i)^2 \leq 0$$

which implies that H_L is negative definite for $L > 1$, and therefore, its minima, constrained to the hypercube $[0,1]^N$ (N_L is the total number of blocks) will always lie in a corner of such hypercube which means that we can use simulated annealing to find the global minimum of H_L, constraining the search to $\{0,1\}^N$. In this case, the energy to be minimized takes the simpler equivalent form (up to an additive constant):

$$U_L = \frac{1}{T_0/L} \sum_{i,j} V(f_L(i), f_L(j)) + \alpha L^2 \sum_i q(f_L(i), g_L(i))$$

The minimum energy solutions for each L can be interpreted as "coarse scale" representations of the original pattern f. Once a solution is obtained, the next refinement (for blocks of size L/2) can be efficiently obtained using the previous solution as a starting point, and initiating the annealing process at a lower temperature.

5. FINAL REMARKS

Due to lack of space we are unable to discuss:

(i) Estimation of boundaries using coupled models on the lattice Z^2 and the dual lattice of bonds on Z^2,

(ii) Estimation of the field and temperature parameter T using the innovations field

(iii) other problems in computation vision such as depth from stero-images, shape from shading etc.

A preliminary account can be found in Marroquin (1985).

Appendix on Simulated Annealing.

Let Ω be a finite set and let $|\Omega|$ denote the cardinality of Ω. Consider the problem of minimizing the energy function:

$$U: \Omega \rightarrow R: i \rightarrow U_i.$$

Let $N_0 = N \ \{0\}$ where N are the natural numbers of $T_k > 0$, $k \in N_0$ be a sequence of real numbers. Consider a Markov chain $\{x_k\}_{k \in N}$ with 1-step transition matrices

$\{P^{(k,k+1)}\}_{k \varepsilon N}$ and some initial distribution constructed on a probability space and let $P_i^{(k)} = P\{x_k = i\}$, $i \varepsilon \Omega$, $k \varepsilon N_0$. The "annealing chain" is simulated as follows:

Suppose $x_k = i$. Then generated a random variable y with $P\{y=j\} = q_{ij}$ where $Q = \{q_{ij}\}_{i,j \varepsilon \Omega}$ a stochastic matrix. Suppose $y=j$, and then define

$$x_{k+1} = \begin{cases} j \text{ if } U_j \leq U_i \\ j \text{ if } U_j > U_i \text{ with probability } e^{(U_i - U_j)/T_k} \\ i \text{ otherwise} \end{cases}$$

We may think of the annealing algorithm as a probabilistic descent algorithm where the Q-matrix represents some prior distribution of "directions", transitions to some or lower energies are always allowed and transitions to higher energies are allowed with positive probability which tends to 0 as $k \to \infty$.

Hajek 1985 has given necessary and sufficient conditions on the rate at which T_k should go to zero such that $P\{x_k \varepsilon S^*\} \to 1$ as $k \to \infty$ where S^* is the set of global minimizing energy states. In this analysis the stochastic matrix Q has to be irreducible and satisfy a weak reversibility condition.

In Gelfand-Mitter 1985 a finite-time analysis of the annealing chain has been performed as well as a result on the rate of convergence of $P\{x_k \varepsilon S^*\} \to 1$ as $k \to \infty$ has been given, under somewhat weaker hypotheses on Q.

Finally, in Tsitsiklis 1985, necessary and sufficient conditions for $P\{x_k \varepsilon S^*\} \to 1$ as $k \to \infty$ are given by considering the annealing chain as a singularly perturbed Markov chain operating under different time-scales (under hypotheses weaker than that of Hajek).

Space does not permit us to give a detailed account of these results.

REFERENCES

1. Carmona, R. (1980): Processes de Diffusion Gouverne par la Forme de Dirichlet de l' Operateur de Schrodinger in Seminaire de Probabilite's XIII, Springer-Verlag, Berlin, New York.

2. Fleming, W. and S.K. Mitter (1982): Optimal Control and Nonlinear Filtering for Nondegenerate Diffusion Processes, Stochastics, 8, pp. 63-78.

3. Geman, S. and D. Geman (1984): "Stochastic Relaxation, Gibbs Distribution, and the Bayesian Restoration of Images". IEEE Trans. Pattern Analysis and Machine Intelligence 6, 721-741 (1984).

4. Gelfand, S. and S.K. Mitter: "Analysis of Simulated Annealing for Optimization", LIDS-P-1494, 1985.

5. Glimm, J. and A. Jaffe (1981): Quantum Physics, Springer-Verlag.

6. Grenander, U., "Tutorial in Pattern Theory", Div. of Applied Math. Brown University (1984).

7. Gross, L. (1972): Existence and Uniqueness of Physical Ground States, J. Functional Analysis, 10, pp. 52-109.

8. Guerra, F., L. Rosen and B. Simon (1975): The $P(\phi)_2$ Euclidean Quantum Field Theory as Classical Statistical Mechanics, Annals of Mathematics, 101, pp. 111-259.

9. Hajek, B.: "Cooling Schedules for Optimal Annealing," Preprint, Dept. Elec. Eng. and Coor. Science Lab., U. Illinois at Champaign-Urbana (1985).

10. Kirkpatrick, S., C.D. Gelatt, and M.P. Vecchi, "Optimization by Simulated Annealing". Science 220 (1983) 671-680.

11. Lasinio, G. Jona, P.K. Mitter (1984): On the Stochastic Quantization of Field Theory, Preprint, LPTHE 84/36, University of Paris VI.

12. Marroquin, J.L. (1985): Probabilistic Solution of Inverse Problems, Doctoral Dissertation, Department of Electrical Engineering and Computer Science, M.I.T.

13. Meyer, P.A. (1978): La Formula de Ito pour le Mouvement Brownian d'apres G. Brosmaler, Lecture Notes in Mathematics, 649, 763-769, Springer-Verlag.

14. Mitter, S.K. (1980): Filtering Theory and Quantum Fields, in Analyse des Systemes, Astreisque, Vol. 75-76, pp. 199-205.

15. Mitter, S.K. (1983): Lectures on Nonlinear Filtering and Stochastic Control in Nonlinear Filtering and Stochastic Control, eds. Mitter, S.K., A. Moro, Springer Lecture Notes in Mathematics, 972, Springer-Verlag, Berlin, New York.

16. Mitter, S.K.: "On the Analogy Between Mathematical Problems of Non-Linear Filtering and Quantum Physics", Ricerche Di Automatica, Vol. 10, No. 2, December 1979.

17. Nelson, E. (1973): Probability Theory and Euclidean Field Theory in : Constructive Quantum Field Theory, G. Velo and A.S. Wightman, eds., Springer-Verlag.

18. Parisi, G. and Wu Yongshi (1981): Perturbation Theory without Gauge Fixing, Scientia Sinica, XXIV, pp. 483-496.

19. Segal, I. (1970): Construction of Nonlinear Local Quantum Processes I, Annals of Mathematics 92, pp. 462-481.

20. Simon, B. (1979): Functional Integration and Quantum Physics, Academic Press, New York.

21. Sinai, Ya G. (1982): Theory of Phase Transitions: Rigorous Results, Pargamon Press, Oxford, New York.

22. Tsitsiklis, J.: "Markov Chains with Rare Transitions and Simulated Annealing", LIDS-P-1497, Laboratory for Information and Decision Systems, M.I.T.

23. Wilson, K.G., "The Renormalization Group: Critical Phenomena and the Kondo

Problem", Rev. Mod. Phys. 47, 4(1973).

Acknowledgement

Research supported by the U.S. Army Res. Off. under Grant DAAG29-84-K-0005 and by the Air Force Office of Scientific Research under grant AFOSR-85-0227.

QUANTUM STOCHASTIC CALCULUS

K.R. Parthasarathy
Indian Statistical Institute
Delhi Centre
7, S.J.S. Sansanwal Marg
New Delhi - 110 016

1. Introduction

The central aim of this lecture is to present to an audience of probability theorists a very brief and somewhat hurried account of some of the recent developments in quantum stochastic calculus which is essentially based on the commutation rules of a free Boson field and Riemann integration. This is simpler than and, in some respects, a generalisation of the classical Ito calculus based on Brownian motion and Poisson process. Furthermore, such an attempt leads to a quantum Ito's formula, Schrödinger and Heisenberg equations in the presence of noise, a canonical isomorphism between the antisymmetric Fock space over $L_2[0,\infty)$ and the Hilbert space of square integrable functions with respect to standard Brownian motion explaining thereby the relations between Fermi, Bose and Wiener chaos and finally a stochastic integral representation of quantum martingales in the symmetric Fock space over $L_2[0,\infty)$.

Most of the work presented here has been done in collaboration with R.L. Hudson. The results of the last section have been obtained in collaboration with K.B. Sinha. Due to the limited scope of a one hour lecture I have not touched on the work of several authors but I hope that the bibliography, although incomplete, is rich enough to convey the growing nature of the subject. Furthermore, the emphasis is more on examples than proofs.

I wish to acknowledge several long and fruitful conversations with R.L. Hudson, R.F. Streater, L.A. Accardi, K.B. Sinha and many other friends on this subject. I also wish to thank Professor T. Hida and the organisers of the Fifteenth Conference on Stochastic Processes and their Applications of the Bernoulli Society for giving me an opportunity to present these ideas and their generous hospitality and financial support.

2. The Boson Fock space and Weyl representation

In this section we introduce the notion of a Boson Fock space and explain the connection with some notions of probability theory.

Let $\mathfrak{h}$ be any complex separable Hilbert space. We shall denote all Hilbert space inner products by $\langle\cdot,\cdot\rangle$ which is antilinear in the first variable. We write

$$\Gamma(\mathfrak{h}) = \mathbb{C} \oplus \mathfrak{h} \oplus \mathfrak{h} \circledS \mathfrak{h} \oplus \ldots \oplus \mathfrak{h}^{\circledS n} \oplus \ldots\ldots$$

where $\mathfrak{h}^{\circledS n}$ is the n-fold symmetric tensor product of $\mathfrak{h}$ and call $\Gamma(\mathfrak{h})$ the symmetric or Boson Fock space over $\mathfrak{h}$. For any $u \in \mathfrak{h}$ we define the coherent or exponential vector $\psi(u)$ in $\Gamma(\mathfrak{h})$ by

$$\psi(u) = 1 \oplus u \oplus \frac{u \otimes u}{\sqrt{2!}} \oplus \ldots \oplus \frac{u^{\otimes n}}{\sqrt{n!}} \oplus \quad \ldots \tag{2.1}$$

where $u^{\otimes n}$ denotes the n-fold product $u \otimes u \otimes \ldots \otimes u$. Observe that

$$\langle\psi(u),\psi(v)\rangle = \exp \langle u,v\rangle. \tag{2.2}$$

We write

$$\Omega = \psi(0) = 1 + 0 + 0 + \ldots$$

and call it the vacuum vector. The subspace $\mathfrak{h}^{\circledS n} \subset \Gamma(\mathfrak{h})$ is called the n-particle subspace and any element in it is called an n-particle vector. Let $\mathcal{U}(\mathfrak{h})$ be the group of all unitary operators on $\mathfrak{h}$ with strong topology and let $\mathcal{E}(\mathfrak{h}) = \mathfrak{h} \circledS \mathcal{U}(\mathfrak{h})$ be the semidirect product of the additive group $\mathfrak{h}$ with its norm topology and $\mathcal{U}(\mathfrak{h})$. As a topological space $\mathcal{E}(\mathfrak{h})$ is the cartesian product of $\mathfrak{h}$ and $\mathcal{U}(\mathfrak{h})$ but the group multiplication is defined by $(u,U)\cdot(v,V) = (u+Uv, UV)$. Then $\mathcal{E}(\mathfrak{h})$ is a topological group called the Euclidean group over $\mathfrak{h}$. With these notations we now state a well known result from quantum theory which plays a fundamental role in our subject.

Theorem 2.1 For any $(u,U) \in \mathcal{E}(\mathfrak{h})$ there exists a unique unitary operator $W(u,U)$ on $\Gamma(\mathfrak{h})$ satisfying

$$W(u,U)\psi(v) = \{\exp(-\tfrac{1}{2}||u||^2 - \langle u,Uv\rangle)\}\psi(Uv+u) \tag{2.3}$$

for all $v \in \mathfrak{h}$. The map $(u,U) \to W(u,U)$ is an irreducible, strongly continuous and projective unitary representation of the group $\mathcal{E}(\mathfrak{h})$. In particular,

$$W(u,U)W(v,V) = (\exp -i\mathrm{Im}\langle u,Uv\rangle)W(u+Uv,UV) \tag{2.4}$$

<u>Remark 2.2</u> The correspondence $u \to W(u,1), u \in \mathfrak{h}$ is a strongly continuous, irreducible and projective unitary representation of the additive group $\mathfrak{h}$ with multiplier $\exp -i\langle u,v\rangle$. The correspondence $U \to \Gamma(U) = W(o,U)$ is a strongly continuous but reducible unitary representation, leaving each n-particle subspace invariant. $\Gamma(U)$ is called the <u>second quantization</u> of U. The map $(u,U) \to W(u,U)$ is called the <u>Weyl representation</u> of $\mathcal{E}(\mathfrak{h})$ in view of the fact that it is an extension of the well known Weyl commutation relations of quantum theory.

Quantum stochastic calculus, as outlined here, depends on the infinitesimal form of the Weyl representation. In order to describe it we introduce a definition. For any $u \in \mathfrak{h}$, $T \in \mathcal{B}(\mathfrak{h})$, the algebra of all bounded operators on $\mathfrak{h}$, define

$$\left.\begin{aligned} a(u)\psi(v) &= \langle u,v\rangle\psi(v), \\ a^{\dagger}(u)\psi(v) &= \frac{d}{d\varepsilon}\psi(v+\varepsilon u)\Big|_{\varepsilon=0}, \\ \lambda(T)\psi(v) &= \frac{d}{d\varepsilon}\psi(e^{\varepsilon T}v)\Big|_{\varepsilon=0}, \end{aligned}\right\} \tag{2.5}$$

where the derivatives are in the strong sense. The exponential vectors generate a dense linear manifold $\mathcal{M}$ in $\Gamma(\mathfrak{h})$ and also constitute a linearly independent set. Thus we can extend $a(u)$, $a^{\dagger}(u)$ and $\lambda(T)$ by linearity to $\mathcal{M}$. Observe that $a(u)$ and $a^{\dagger}(u)$ are adjoint to each other on $\mathcal{M}$. If $T^{\dagger}$ denotes the adjoint of T then $\lambda(T)$ and $\lambda(T^{\dagger})$ are adjoint to each other. Thus $a(u)$, $a^{\dagger}(u)$ and $\lambda(T)$ are closable. We close them and denote them by the same symbols. Then finite particle vectors belong to their domains. $a(u)$ maps an n-particle vector into an (n-1)-particle vector and $a(u)\Omega = 0$. $a^{\dagger}(u)$ takes an n-particle vector to (n+1)-particle vector whereas $\lambda(T)$ leaves the n-particle subspace invariant. In view of these properties we call $a(u)$, $a^{\dagger}(u)$ and $\lambda(T)$ respectively <u>annihilation</u>, <u>creation</u> and <u>conservation</u> operators. They obey the following commutation relations on $\mathcal{M}$:

$$\left.\begin{aligned} [a(u), a(v)] &= [a^{\dagger}(u), a^{\dagger}(v)] = 0, \\ [a(u), a^{\dagger}(v)] &= \langle u,v\rangle, \ [\lambda(T_1), \lambda(T_2)] = \lambda([T_1,T_2]), \\ [\lambda(T), a(u)] &= -a(T^{\dagger}u), \\ [\lambda(T), a^{\dagger}(v)] &= a^{\dagger}(Tv). \end{aligned}\right\} \tag{2.6a}$$

Furthermore,, the map $u \to a(u)|_{\mathcal{M}}$ is antilinear whereas the map $T \to \lambda(T)|_{\mathcal{M}}$

is linear. Our next result shows that the commutation relations (2.6a) constitute an infinitesimal description of the Weyl representation:

Theorem 2.3 On the linear manifold $\mathcal{M}$ generated by the exponential vectors

$$W(u,U) = (\exp - \tfrac{1}{2}||u||^2)\, e^{a^\dagger(u)}\Gamma(U)e^{-a(Uu)}. \tag{2.6}$$

If $U_s = \exp isH$ where H is a bounded selfadjoint operator on $\mathfrak{h}$ then

$$\Gamma(U_s) = \exp is\lambda(H) \quad \text{for all} \quad s \in \mathbb{R}. \tag{2.7}$$

We now proceed to explain the relation between the Weyl representation and the infinitely divisible distributions of probability theory. To this end let μ be a Lévy measure on $\mathbb{R}$ so that μ is σ-finite, $\mu(\{x: |x| > \varepsilon\}) < \infty$ for each $\varepsilon > 0$ and

$$\int_{|x|<\delta} x^2\mu(dx) < \infty \quad \text{for some} \quad \delta > 0.$$

Let $\mathfrak{h} = L_2(\mu)$. Define the unitary representation U of $\mathbb{R}$ in $\mathfrak{h}$ by

$$(U_x f)(y) = e^{ixy} f(y), \; f \in L_2(\mu).$$

Let

$$\eta_x(y) = \begin{cases} x & \text{if } y = 0, \\ e^{ixy} - 1 & \text{if } y \neq 0. \end{cases}$$

Then $\eta_x \in L_2(\mu)$ for each x. Then we have the following result.

Theorem 2.4 (R.F. Streater [26], H.Araki [4], K.R. Parthasarathy and K. Schmidt [20]). Let

$$\tilde{W}_x = \{\exp i\int(\sin xy - \frac{xy}{1+y^2})\mu(dy)\}W(\eta_x, U_x)$$

where W on the right hand side denotes the Weyl representation of $\mathcal{E}(L_2(\mu))$. Then the correspondence $x \to \tilde{W}_x$ is a unitary representation of $\mathbb{R}$ in $\Gamma(L_2(\mu))$ and

$$\langle\Omega, \tilde{W}_x\Omega\rangle = \exp \int_{\mathbb{R}\setminus\{0\}} (e^{ixy}-1-\frac{ixy}{1+y^2})\mu(dy)-\mu(\{0\})x^2/2 . \tag{2.8}$$

In quantum theory a real valued random variable X on a probability space is identified with the selfadjoint operator of multiplication by

X in the Hilbert space of square integrable functions with respect to the underlying probability measure. More generally a classical stochastic process $\{X_t\}$ is to be interpreted as the set $\{X_t\}$ of the respective mutually commuting selfadjoint multiplication operators. In Theorem 2.4 if $\tilde{W}_x = \exp ixX$ and the selfadjoint operator X has the spectral resolution $X = \int_{\mathbb{R}} \lambda \, P(d\lambda)$ then the probability measure $\langle\Omega, P(E)\Omega\rangle$ has characteristic function given by the right hand side of (2.8). Thus all infinitely divisible distributions can be realised as selfadjoint operators in a Fock space. Replacing $L_2(\mu)$ by $L_2(\mu) \otimes L_2([0,\infty))$ it is possible to realise a stochastic process with independent increments as a one parameter family of commuting selfadjoint operators in $\Gamma(L_2(\mu) \otimes L_2[0,\infty))$. We may replace $L_2(\mu)$ by an abstract infinite dimensional Hilbert space $\mathfrak{h}$ and vary instead the pair η, U in Theorem 2.4 and realise every process with independent increments in a single Fock space $\Gamma(\mathfrak{h} \otimes L_2[0,\infty))$. However, selfadjoint operators representing two different classical processes need not commute.

We now pay special attention to the standard Brownian motion and state a theorem arising from the well known works of N. Wiener, K. Ito and I. Segal.

<u>Theorem 2.5</u> Let $\mathfrak{h} = L_2([0,\infty))$ and let P denote the Wiener probability measure of the standard Brownian motion in the time interval $[0,\infty)$. Then there exists a unique unitary isomorphism $\Theta : \Gamma(\mathfrak{h}) \to L_2(P)$

$$\Theta\psi(u) = \exp \int_0^\infty u \, dw - \tfrac{1}{2} \int_0^\infty u^2 \, dt \quad \text{for all} \quad u \in \mathfrak{h} \tag{2.9}$$

where w denotes a typical Wiener path. Furthermore, for every real $u \in \mathfrak{h}$, $f \in L_2(P)$

$$(\Theta\{a(u)+a^\dagger(u)\}^\sim \Theta^{-1} f)(w) = f(w) \int_0^\infty u \, dw \tag{2.10}$$

$$(\Theta \exp \tfrac{1}{2}(a(u)-a^\dagger(u))^\sim \Theta^{-1} f)(w) = f(\tau_u w) \exp(-\tfrac{1}{2}\int_0^\infty u \, dw - \tfrac{1}{4}\int_0^\infty u^2 dt) \tag{2.11}$$

where

$$(\tau_u w)(t) = w(t) + \int_0^t u(s) dw(s)$$

and $\sim$ denotes closure.

<u>Remark 2.6</u> Let $q(u) = (a(u) + a^\dagger(u))^\sim$, $p(v) = -\tfrac{1}{2} i \, (a(v) - a^\dagger(v))^\sim$. Then $q(u)$, $p(v)$ are selfadjoint operators obeying the well known Heisenberg

commutation relations for a free Boson field :

$$[q(u),\ q(v)] = [p(u),\ p(v)] = 0$$

$$[q(u),\ p(v)] = i\langle u,v\rangle \quad \text{for all} \quad u,v \in \mathfrak{h} .$$

Theorem 2.5 tells us how to realise these operators in $L_2(P)$, when $\mathfrak{h} = L_2[0,\infty)$.

Define $Q(t) = q(\chi_{[0,t]})$, χ denoting indicator. Then $\{Q(t), t \geq 0\}$ is a commuting family of selfadjoint operators which appear as multiplication operators by $\{w(t),\ t \geq 0\}$ in $L_2(P)$. In other words we have realised the standard Brownian motion through the operators $\{Q(t)\}$ in the state Ω in $\Gamma(L_2[0,\infty))$.

In the light of the above remarks the following questions arise: (1) how do stochastic integrals, Ito's formula and diffusions look like in the Fock space $\Gamma(L_2[0,\infty))$? (2) What is the role of conservation operators in constructing processes with independent increments and point processes in the light of Theorems 2.4 and 2.5? (3) What is the relation between Fermion field operators over $\mathfrak{h} = L_2([0,\infty))$ and standard Brownian motion? We shall offer a few glimpses on some of these aspects in what follows.

3. Integration and quantum Ito's formula

We shall now present a brief summary of quantum stochastic integration in the special case of the Hilbert space $\mathfrak{h}_0 \otimes \Gamma(L_2[0,\infty))$ where $\mathfrak{h}_0$ is a fixed Hilbert space called the <u>initial space</u>. We write

$$H = \mathfrak{h}_0 \otimes \Gamma(L_2[0,\infty)), \qquad H_t = \mathfrak{h}_0 \otimes \Gamma(L_2[0,t])$$

$$H^t = \Gamma(L_2[t,\infty)),\ \Omega^t = \text{vacuum vector in } H^t.$$

Note that $H = H_t \otimes H^t$ via the identification

$$u \otimes \psi(f) = \{u \otimes \psi(f|_{[0,t]})\} \otimes \psi(f|_{[t,\infty)})$$

where $u \in \mathfrak{h}_0$, $f \in L_2[0,\infty)$. The Hilbert space H_t is to be interpreted as the space of states describing events occurring in the time interval $[0,t]$. We now introduce the key notion of an adapted process of operators in H. Roughly speaking, a family of operators $X = \{X(t),\ t \geq 0\}$ is called an <u>adapted process</u> if $X(t) = X_0(t) \otimes 1^t$ for each t where $X_0(t)$ is an operator in H_t and 1^t is the identity operator in H^t. However, we introduce a formal definition. Let $\mathcal{D}$ be a dense linear manifold in the initial space $\mathfrak{h}_0$. We say that X is an <u>adapted process with initial domain $\mathcal{D}$</u> if for all $u \in \mathcal{D}$, $f \in L_2[0,\infty)$

(i) $u \otimes \psi(f) \in \mathcal{D}(X(t))$, the domain of $X(t)$;

(ii) $X(t)\, u \otimes \psi(f\chi_{[0,t]})$ is of the form $\xi(t) \otimes \Omega^t$ where $\xi(t) \in H_t$ and $X(t)u \otimes \psi(f) = \xi(t) \otimes \psi(f|_{[t,\infty)})$. The adapted process X is called <u>bounded</u> if each $X(t)$ is a bounded operator. It is called <u>Hilbert-Schmidt</u> if $X(t) = X_o(t) \otimes 1^t$ where $X_o(t)$ is a Hilbert-Schmidt operator in H_t for each t. It is called a <u>martingale</u> if

$$\langle u \otimes \psi(f\chi_{[0,s]}),\ X(t)v \otimes \psi(g\chi_{[0,s]})\rangle =$$

$$\langle u \otimes \psi(f\chi_{[0,s]}),\ X(s)v \otimes \psi(g\chi_{[0,s]})\rangle$$

for all $u, v \in \mathcal{D}$, $f,g \in L_2[0,\infty)$, $s < t$. It is called <u>simple</u> if there exist $0 < t_1 < t_2 < \ldots$, $\lim_{n \to \infty} t_n = \infty$ and

$$X(t) = X(0)\chi_{[0,t_1)} + X(t_1)\chi_{[t_1,t_2)} + \ldots + X(t_j)\chi_{[t_j,t_{j+1})} + \ldots \tag{3.1}$$

We now introduce three basic processes with initial domain $\mathfrak{h}_o$:

$$\left.\begin{aligned} A(t) &= 1_o \otimes a(\chi_{[0,t]}), \\ A^\dagger(t) &= 1_o \otimes a^\dagger(\chi_{[0,t]}), \\ \Lambda(t) &= 1_o \otimes \lambda(\chi_{[0,t]}) \end{aligned}\right\} \tag{3.2}$$

where 1_o is identity in $\mathfrak{h}_o$, a, $a^\dagger$, λ are defined by (2.5), $\chi_{[0,t]}$ is the indicator function of the interval $[0,t]$ considered as an element of $L_2[0,\infty)$ in the definitions of A and $A^\dagger$ and as the projection operator of multiplication in the definition of Λ. The processes A, $A^\dagger$ and Λ are all martingales. We call them <u>annihilation</u>, <u>creation</u> and <u>conservation</u> <u>martingale</u> respectively in view of the remarks in § 2.

For any simple adapted process with initial domain $\mathcal{D}$ and satisfying (3.1) we define

$$\int_0^t X(s)\,dM(s) = \sum_{j=0}^{\infty} X(t_j)\{M(t_{j+1} \wedge t) - M(t_j \wedge t)\} \tag{3.3}$$

where $s \wedge t$ denotes the minimum of s and t, $t_o = 0$ and M is any one of the martingales A, $A^\dagger$ and Λ. The domain of the operator (3.3) includes all vectors of the form $u \otimes \psi(f)$. Furthermore $Y(t) = \int_0^t X(s)\,dM(s)$ is a martingale with initial domain $\mathcal{D}$. Let

$$\int_0^t X(s)\,ds = \sum_{j=0}^{\infty} X(t_j)(t_{j+1} \wedge t - t_j \wedge t) \tag{3.4}$$

denote the usual Riemann sum. The operator $X(t_j)$ acts essentially in H_{t_j} whereas $M(t_{j+1} \wedge t) - M(t_j \wedge t)$ acts essentially in $\Gamma(L_2[t_j, t_{j+1}])$ and hence the two commute on the domain $\mathcal{D} \underline{\otimes} \mathcal{M}$, the algebraic tensor product of $\mathcal{D}$ and the linear manifold generated by exponential vectors. Hence we restrict the definitions (3.3) and (3.4) to the domain $\mathcal{D} \underline{\otimes} \mathcal{M}$ and do not distinguish between the processes $\int_0^t X(s)dM(s)$ and $\int_0^t (dM(s))X(s)$ when $M = A^\dagger$, A or Λ.

Let $\underline{E}(t) = (E_j(t),\ t \geq 0,\ 1 \leq j \leq 4)$ be four simple adapted processes with initial domain $\mathcal{D}$. Define the integral $\int_0^t E_1 d\Lambda + E_2 dA + E_3 dA^\dagger + E_4 ds$ to be the sum of the four individual integrals with respect to $d\Lambda$, dA, $dA^\dagger$ and ds on $\mathcal{D} \underline{\otimes} \mathcal{M}$. Using definition (2.5), the commutation relations (2.6a) and the property that $d\Lambda$, dA, $dA^\dagger$ commute with E_j's we obtain by an elementary but tedious computation the following fundamental lemma:

<u>Lemma 3.1</u> Let E_i, F_i, $1 \leq i \leq 4$ be simple adapted processes with initial domain $\mathcal{D}$ and let

$$X(t) = \int_0^t E_1 d\Lambda + E_2 dA + E_3 dA^\dagger + E_4 ds,\quad Y(t) = \int_0^t F_1 d\Lambda + F_2 dA + F_3 dA^\dagger + F_4 ds. \qquad (3.5)$$

Then for all $u, v \in \mathcal{D}$, $f, g \in L_2[0,\infty)$ we have

$$\langle u \otimes \psi(f),\ Y(t)v \otimes \psi(g)\rangle =$$
$$\int_0^t \langle u \otimes \psi(f),\ (\bar{f}(s)g(s)F_1(s) + g(s)F_2(s) + \bar{f}(s)F_3(s) + F_4(s))v \otimes \psi(g)\rangle ds, \qquad (3.6)$$

$$\langle X(t)u \otimes \psi(f),\ Y(t)v \otimes \psi(g)\rangle =$$
$$\int_0^t \{\langle X(s)u \otimes \psi(f), (\bar{f}(s)g(s)F_1(s) + g(s)F_2(s) + \bar{f}(s)F_3(s) + F_4(s))v \otimes \psi(g)\rangle$$
$$+ \langle (f(s)\bar{g}(s)E_1(s) + f(s)E_2(s) + \bar{g}(s)E_3(s) + E_4(s))u \otimes \psi(f),\ Y(s)v \otimes \psi(g)\rangle$$
$$+ \bar{f}(s)g(s)\langle E_1(s)u \otimes \psi(f),\ F_1(s)v \otimes \psi(g)\rangle + \bar{f}(s)\langle E_1(s)u \otimes \psi(f),\ F_3(s)v \otimes \psi(g)\rangle$$
$$+ g(s)\langle E_3(s)u \otimes \psi(f),\ F_1(s)v \otimes \psi(g)\rangle + \langle E_3(s)u \otimes \psi(f),\ F_3(s)v \otimes \psi(g)\rangle\} ds \qquad (3.7)$$

Putting $E_i = F_i$, $u = v$, $f = g$ in the above lemma we get

<u>Corollary 3.2</u> Let $X(t)$ be as in (3.5). Then

$$||X(t)u \otimes \psi(f)||^2 = \int_0^t \{2\,\mathrm{Re}\langle X(s)u \otimes \psi(f), (|f(s)|^2 E_1(s) + f(s)E_2(s) +$$
$$+ \bar{f}(s)E_3(s) + E_4(s))u \otimes \psi(f)\rangle + ||(f(s)E(s) + G(s))u \otimes \psi(f)||^2\} ds \qquad (3.8)$$

for all $u \in \mathcal{D}$, $f \in L_2[0,\infty)$.

An argument similar to the proof of the classical Gronwall's inequality but using (3.8) leads to the following inequality.

Corollary 3.3 Let $X(t)$ be as in (3.5). Then

$$||X(t)u \otimes \psi(f)||^2 \leq 3\int_0^t \exp\{(t-s+3\int_s^t |f(\tau)|^2 d\tau)\}(||f(s)E_1(s)u \otimes \psi(f)||^2 + \sum_{j=2}^{4} ||E_j(s)u \otimes \psi(f)||^2)ds \tag{3.9}$$

for all $u \in \mathcal{D}$, $f \in L_2[0,\infty)$.

Inequality (3.9) enables us to complete the definition of the integral (3.5) for simple adapted processes to the class of all quadruples $\underline{E} = (E_i(t),\ 1 \leq i \leq 4)$ of adapted processes with initial domain $\mathcal{D}$ obeying the condition

$$\int_0^t (||f(s)E_1(s)u \otimes \psi(f)||^2 + \sum_{j=2}^{4} ||E_j(s)u \otimes \psi(f)||^2)ds < \infty$$

for all $u \in \mathcal{D}$, $f \in L_2[0,\infty)$, $0 < t < \infty$. We shall denote this class by $\mathcal{A}_2(\mathcal{D})$. In view of the completion procedure it follows immediately that for $\underline{E}$, $\underline{F} \in \mathcal{A}_2(\mathcal{D})$ the integrals (3.5) are defined on $\mathcal{D} \underline{\otimes} \mathcal{M}$. Furthermore $X = \{X(t)\}$ and $Y = \{Y(t)\}$ are adapted processes with initial domain $\mathcal{D}$ and the identities (3.6) and (3.7) hold. If $E_4 = 0$ then X is a martingale.

If we look at (3.6) and (3.7) it is clear that the first two inner products on the right hand side of (3.7) account for a classical integration by parts formula but all the remaining four terms introduce a <u>correction</u>. From this point of view (3.7) can be interpreted as a quantum Ito's formula.

Suppose now that X, Y are adapted processes with initial domain $\mathcal{D}$ and

$$X(t) = X(0) + \int_0^t E_1 d\Lambda + E_2 dA + E_3 dA^\dagger + E_4\, ds,$$

$$Y(t) = Y(0) + \int_0^t F_1 d\Lambda + F_2 dA + F_3 dA^\dagger + F_4 ds,$$

where $\underline{E}$ and $\underline{F}$ belong to $\mathcal{A}_2(\mathcal{D})$. Then we write

$$dX = E_1 d\Lambda + E_2 dA + E_3 dA^\dagger + E_4\, dt,$$

$$dY = F_1 d\Lambda + F_2 dA + F_3 dA^\dagger + F_4\, dt.$$

Suppose that $X(t)$, $Y(t)$, $E_j(t)$, $F_j(t)$ $(1 \leq j \leq 4)$ are all bounded operators such that

$$\sup_{0\leq s\leq t} \max\{||X(s)||, ||Y(s)||, ||E_j(s)||, ||F_j(s)|| \quad 1\leq j \leq 4\} < \infty .$$

(In the next section we shall meet many such examples.) Then (3.7) can be rewritten as

$$d(X^\dagger(t)Y(t)) = (dX^\dagger(t))Y(t)+X^\dagger(t)dY(t)+dX^\dagger(t)\cdot dY(t) \tag{3.10}$$

where

$$dX^\dagger = E_1^\dagger\, d\Lambda + E_3^\dagger\, dA + E_2^\dagger\, dA^\dagger + E_4^\dagger\, dt,$$

$X^\dagger$, $E_j^\dagger$ are adjoints of X and E_j, the basic differentials $d\Lambda, dA, dA^\dagger$ and dt commute with adapted processes and the <u>correction term</u> $dX^\dagger\cdot dY$ is evaluated by combining this with extension by bilinearity of the multiplication table

	$d\Lambda$	dA	$dA^\dagger$	dt
$d\Lambda$	$d\Lambda$	0	$dA^\dagger$	0
$dA^\dagger$	0	0	0	0
dA	dA	0	dt	0
dt	0	0	0	0

(3.11)

From Remark 2.6 it follows that $Q(t) = A(t) + A^\dagger(t)$ is nothing but standard Brownian motion. From (3.11) we have $dQ.dQ = dt$ which, in the vacuum state, is the classical Ito product formula for the differential of Brownian motion.

It turns out that the classical Poisson process of intensity ν can be identified in $\Gamma(L_2[0,\infty))$ with the adapted process

$$N_\nu(t) = \Lambda(t) + \nu^{\frac{1}{2}}(A(t) + A^\dagger(t)) + \nu t. \tag{3.12}$$

It follows from (3.11) that $dN_\nu\cdot dN_\nu = dN_\nu$ which is the classical Ito's product formula for the differential of a Poisson process. It is also interesting to note that (3.12) leads to a "quantum central limit theorem".

$$\lim_{\nu\to\infty} \frac{N_\nu(t)-\nu t}{\sqrt{\nu}} = Q(t) .$$

4. Examples and Applications

(a) Let the initial space $\mathfrak{h}_o = \mathbb{C}$ and let

$$W(t) = W(f\chi_{[0,t]}, e^{i\theta\chi_{[0,t]}})$$

be the Weyl operator defined by Theorem 2.1 where θ is a real valued Borel function on $[0,\infty)$, $e^{i\theta\chi_{[0,t]}}$ is the unitary operator of multiplication by the function $e^{i\theta\chi_{[0,t]}}$ in $L_2[0,\infty)$ and $f \in L_2[0,\infty)$. Then $W = \{W(t)\}$ is a unitary operator valued adapted process. It follows from Theorem 2.3 that

$$dW = \{(e^{i\theta}-1)d\Lambda + f dA^{\dagger} - \bar{f} e^{i\theta} dA - \tfrac{1}{2}|f|^2 dt\}W.$$

In particular, W obeys a "quantum stochastic differential equation". If we write

$$X(t) = (\exp \tfrac{1}{2} \int_o^t |f|^2 \, ds)\, W(t)$$

then X is a bounded martingale satisfying

$$dX = \{(e^{i\theta}-1)d\Lambda + f dA^{\dagger} - \bar{f}e^{i\theta} \, dA\}X.$$

(b) Let $\mathfrak{h}_o$ be arbitrary and let R,L,H be bounded operators on $\mathfrak{h}_o$ such that R is unitary and H is selfadjoint. We now consider the constant adapted processes $R(t) = R \otimes 1$, $L(t) = L \otimes 1$, $H(t) = H \otimes 1$ for all t where 1 denotes the identity in $\Gamma(L_2[0,\infty))$. It is a theorem that there exists a unique unitary operator valued adapted process $U = \{U(t)\}$ obeying the quantum stochastic differential equation

$$U(o) = 1, \quad dU = \{(R-1)d\Lambda + L dA^{\dagger} - L^{\dagger}R dA - (iH + \tfrac{1}{2}L^{\dagger}L)dt\}U \qquad (4.1)$$

The above equation may be solved by the classical Cauchy-Picard iterative procedure as outlined in [13]. If R = 1, L = 0, (4.1) reduces to the abstract Schrödinger equation $dU = -iHU dt$. Thus we may view (4.1) as a Schrödinger equation in the presence of "noise" or a quantum Langevin equation in the Schrödinger picture. We may also write (4.1) as

$$\frac{dU}{dt} = -i(H + i\{(R-1)\frac{d\Lambda}{dt} + L\frac{dA^{\dagger}}{dt} - L^{\dagger}R\frac{dA}{dt} - \tfrac{1}{2}L^{\dagger}L\})$$

where the term within wavy brackets is a "singular" time dependent energy perturbation. The so called Wigner-Weisskopf atom is described by a similar Schrödinger equation [27].

If X is a bounded operator in the initial space $\mathfrak{h}_o$ which is interpreted as $X \otimes 1$ in $\mathfrak{h}_0 \otimes \Gamma(L_2[0,\infty))$ and U satisfies (4.1) then by quantum Ito's formula

$$d(U^\dagger XU) = U^\dagger\{(R^\dagger XR-1)d\Lambda + R^\dagger [X,L]dA^\dagger + [L^\dagger,X]RdA + \mathcal{L}(X)dt\}U \quad (4.2)$$

where

$$\mathcal{L}(X) = i[H,X] - \tfrac{1}{2}(L^\dagger LX + XL^\dagger L) + L^\dagger XL. \quad (4.3)$$

Equation (4.2) describes the evolution of any observable X under the influence of a unitary evolution U which may be interpreted as a quantum diffusion with drift H, position-momentum diffusion coefficient L and a rotational diffusion coefficient R.

There are interesting examples where H and L are unbounded operators but they are outside the scope of this brief survey. We can also replace the Fock space $\Gamma(L_2[0,\infty))$ by $(L_2[0,\infty) \otimes k)$ where $\dim k > 1$ and study diffusions where the noise has a finite or infinite number of degrees of freedom.

It is of great interest to note that the coefficient of dt in the Heisenberg picture (4.2) is given by (4.3) where the map $X \to \mathcal{L}(X)$ turns out to be the generator of a contraction semigroup $\{e^{t\mathcal{L}}, t \geq 0\}$ of completely positive maps on the $*$-algebra $\mathcal{B}(\mathfrak{h})$. Such semigroups describe the socalled quantum Markov processes [8], [10], [19].

(c) We shall now illustrate a quantum diffusion with two degrees of freedom which can be described in terms of two independent standard Brownian motions w_1 and w_2. Denoting Wiener measure by P we consider the Hilbert space $\tilde{\mathcal{H}} = L_2(\mathbb{R}) \otimes L_2(P \times P)$. Let (q,p) be the canonical Schrödinger pair of position and momentum operators in $L_2(\mathbb{R})$. Define the unitary operator $U(t)$ as multiplication by

$$U(t) = e^{-i\int_0^t w_1(s)dw_2(s)}\, e^{ipw_1(t)}\, e^{iqw_2(t)}$$

with $\tilde{\mathcal{H}}$ being interpreted as $L_2(\mathbb{R})$-valued random variables on the probability space of $P \times P$. Then

$$U(0) = 1,\ dU = \{ipdw_1 + iqdw_2 - \tfrac{1}{2}(p^2+q^2)dt\}U.$$

If X is a bounded operator in $L_2(\mathbb{R})$ then

$$\mathbb{E}\, U(t)^\dagger (X \otimes 1)\, U(t) = e^{t\mathcal{L}}(X)$$

where

$$\mathcal{L}(X) = \{-\tfrac{1}{2}(p^2X + Xp^2) + pXp\} + \{-\tfrac{1}{2}(q^2X+Xq^2)+qXq\}$$

and $\mathbb{E}$ denotes expectation with respect to $P \times P$.

In particular

$$\mathcal{L}(\phi(q)) = \tfrac{1}{2}\phi''(q), \quad \mathcal{L}(\phi(p)) = \tfrac{1}{2}\phi''(p)$$

for any twice differentiable ϕ with ϕ, ϕ' and ϕ'' bounded and continuous. This shows that under the quantum diffusion driven by U with diffusion coefficients p, q for the Brownian noise differentials dw_1, dw_2 and drift coefficient = 0, postion and momentum execute standard Brownian motions which are out of phase.

(d) We now present an example of a "spin diffusion" under a classical Poisson noise. Let $\mathfrak{h}_o = \mathbb{C}^2$ and let $\sigma_1, \sigma_2, \sigma_3$ be the Pauli spin matrices. Suppose N_j, $j = 1,2,3$ are three independent Poisson processes with intensities ν_j, $j = 1,2,3$ respectively. Define

$$U(t) = (-1)^{\int_o^t N_1 dN_2 + \int_o^t (N_1+N_2)dN_3} \sigma_1^{N_1(t)} \sigma_2^{N_2(t)} \sigma_3^{N_3(t)} \tag{4.4}$$

in $\mathbb{C}^2 \otimes L^2(N_1, N_2, N_3)$. i.e., U(t) is multiplication by the 2 x 2 random matrix on the right hand side of (4.4). For any $X \in \mathcal{B}(\mathbb{C}^2)$ we have

$$\mathbb{E}\, U(t)^\dagger (X \otimes 1) U(t) = e^{t\mathcal{L}}(X)$$

where

$$\mathcal{L}(X) = \sum_{j=1}^{3} \nu_j (\sigma_j^\dagger X \sigma_j - X)$$

and $\mathbb{E}$ denotes expectation with respect to the probability measure of (N_1, N_2, N_3). Thus the unitary evolution U drives a "spin diffusion" with diffusion coefficients $\sqrt{\nu_j}\,\sigma_j$, $j = 1,2,3$ and drift = 0.

Examples (c) and (d) show that when the diffusion coefficients are selfadjoint (like p, q, σ_j etc.) one has the possibility of achieving the diffusion with classical noise but not otherwise.

(e) Let $\mathfrak{h}_o = \mathbb{C}$. For any real θ let $Q_\theta(t)$ denote the closure of $e^{-i\theta} A(t) + e^{i\theta} A^\dagger(t)$. Then $Q_\theta(t)$ is selfadjoint and

$$[Q_{\theta_1}(t_1), Q_{\theta_2}(t_2)] = 2i[\sin(\theta_2-\theta_1)]\, t_1 \wedge t_2.$$

For each θ, Q_θ is a standard Brownian motion in the vacuum state Ω. For $0 \leq \theta_1 < \theta_2 < \pi$, Q_{θ_1} and Q_{θ_2} do not commute. But

$$\Gamma(e^{i\theta})\, Q_o(t)\, (e^{-i\theta}) = Q_\theta(t)$$

and $\Gamma(e^{i\theta})\Omega = \Omega$. In other words the different standard Brownian motions are "out of phase". In differential notation

$$dQ_\theta = e^{-i\theta}\, dA + e^{i\theta}\, dA^\dagger.$$

(f) Let $\mathfrak{h}_0 = \mathbb{C}$, $J(t) = \Gamma(e^{i\pi\chi_{[0,t]}})$ in $\Gamma(L_2[0,\infty))$. Then $J(t)$ is unitary and selfadjoint for each t. Furthermore J is a martingale satisfying the equation

$$dJ = -2\, J d\Lambda,\ J(0) = 1.$$

We write

$$F(f) = \int_0^\infty \bar{f}(s)J(s)dA(s),\quad F^\dagger(f) = \int_0^\infty f(s)J(s)dA^\dagger(s). \tag{4.5}$$

As a consequence of Ito's formula it turns out that

$$\begin{aligned} F(f)F(g) + F(g)F(f) &= 0 \\ F(f)F^\dagger(g) + F^\dagger(g)F(f) &= \langle f,g\rangle \\ \Gamma(e^{i\theta})F(f)\Gamma(e^{-i\theta}) &= F(e^{i\theta}f) \end{aligned} \tag{4.6}$$

on the domain $\mathcal{M}$. The fact that $F(f)$ and $F^\dagger(f)$ are adjoint to each other and the middle equation in (4.6) imply that $F(f)$ and $F^\dagger(f)$ extend themselves to bounded operators of norm $\le ||f||$, $F^\dagger(f)$ is the adjoint of $F(f)$ and the relations (4.6) hold everywhere. Furthermore $F(f)\Omega = 0$ and the family

$$\{F^\dagger(f_1)\ldots F^\dagger(f_n)\Omega,\ f_j \in L_2[0,\infty),\ 1 \le j \le n,\ n = 1,2,\ldots\}$$

is total in $\Gamma(L_2[0,\infty))$. Thus we have realised through (4.5) an irreducible representation of the Fermion field in a Boson Fock space!

For any Hilbert space $\mathfrak{h}$ introduce the skew symmetric or Fermion Fock space

$$\Gamma_-(\mathfrak{h}) = \mathbb{C} \oplus \mathfrak{h} \oplus \mathfrak{h}\wedge\mathfrak{h} \oplus \ldots \oplus \mathfrak{h}^{\wedge^n} \oplus \ldots$$

with vacuum Ω_- defined by

$$\Omega_- = 1 \oplus 0 \oplus 0 \oplus \ldots$$

where $\mathfrak{h}^{\wedge^n}$ denotes the n-fold skew symmetric tensor product of $\mathfrak{h}$. For any $u \in \mathfrak{h}$ define Fermion annihilation and creation operators on $\Gamma_-(\mathfrak{h})$ by

$$\begin{aligned} a_-(u)v_1\wedge\ldots\wedge v_n &= n^{-\frac{1}{2}} \sum_{j=1}^{n} \langle u,v_j\rangle(-1)^{j-1}\, v_1\wedge\ldots\wedge\hat{v}_j\wedge\ldots\wedge v_n, \\ a_-^\dagger(u)v_1\wedge\ldots\wedge v_n &= (n+1)^{\frac{1}{2}}\, u\wedge v_1\wedge\ldots\wedge v_n, \end{aligned}$$

Λ denoting skew symmetric tensor product and $\hat{}$ over v_j means its omission. The family $\{a_-(u), a_-^\dagger(v), u,v \in \mathfrak{h}\}$ is an irreducible representation of the Fermion field over $\mathfrak{h}$ with vacuum Ω_-. Using the uniqueness theorem for such representations we obtain the following result.

Theorem 4.1 There exists a unique unitary isomorphism

$$\Phi : \Gamma_-(L_2[0,\infty)) \to \Gamma(L_2[0,\infty))$$

such that

$$\Phi\Omega_- = \Omega$$

$$\Phi\{a_-^\dagger(f_1)\dots a_-^\dagger(f_n)\Omega_-\} = \int_{0<s_1<s_2\dots<s_n<\infty} \det((f_i(s_j))dA^\dagger(s_1)\dots dA^\dagger(s_n)\Omega \tag{4.7}$$

for all $f_1,\dots,f_n \in L_2[0,\infty)$, $n = 1,2,\dots$.

Corollary 4.2 Let Θ be the unitary isomorphism defined by Theorem 2.5. Then $\Theta \circ \Phi$ is a unitary isomorphism from the skew-symmetric or Fermion Fock space $\Gamma_-(L_2[0,\infty))$ to $L_2(P)$, P being Wiener measure. Furthermore

$$\Theta \circ \Phi\{a_-^\dagger(f_1)\dots a_-^\dagger(f_n)\Omega_-\} = \int_{0<s_1<s_2<\dots<s_n<\infty} \det((f_i(s_j))) dw(s_1)\dots dw(s_n) \tag{4.8}$$

for all $f_j \in L_2[0,\infty)$, $1 \leq j \leq n$, $n = 1,2,\dots$, where the right hand side is a multiple Wiener-Ito integral.

Remark 4.3 If we write

$$F(t) = \int_0^t \Gamma(e^{i\pi\chi_{[0,t]}})\, dA(s)$$

$$F^\dagger(t) = \int_0^t \Gamma(e^{i\pi\chi_{[0,s]}})\, dA^\dagger(s)$$

for the Fermion annihilation and creation processes then

$$a^\dagger(f_1)a^\dagger(f_2)\dots a^\dagger(f_n)\Omega = \int_{0<s_1<\dots<s_n<\infty} \mathrm{per}((f_i(s_j)))dF^\dagger(s_1)\dots dF^\dagger(s_n)\Omega \tag{4.9}$$

where $a^\dagger(f)$ is the Boson creation operator and "per" denotes the permanent of a matrix. Thus (4.7), (4.8) and (4.9) imply that "Boson, Fermion and Wiener Chaos" can be expressed in terms of each other.

5. The martingale representation theorem in $\Gamma(L_2[0,\infty))$

Let $\underline{E} = (E_i, 1 \leq i \leq 4)$ be four adapted processes in $\Gamma(L_2[0,\infty))$ satisfying the following properties: (i) for fixed $f,g \in L_2[0,\infty)$ the map $t \to \langle\psi(f), E_i(t)\psi(g)\rangle$ is continuous; (ii) for fixed t the map $g \to E_i(t)\psi(g)$ is continuous; (iii) $E_1 d\Lambda + E_2 dA + E_3 dA^\dagger + E_4 dt = 0$. Then we claim that E_i vanishes on the linear manifold $\mathcal{M}$ generated by the exponential vectors for each i. Indeed, from Lemma 3.1 we conclude that for each $f,g \in L_2[0,\infty)$

$$\bar{f}(t)g(t)\langle\psi(f), E_1(t)\psi(g)\rangle + g(t)\langle\psi(f), E_2(t)\psi(g)\rangle$$
$$+ \bar{f}(t)\langle\psi(f), E_3(t)\psi(g)\rangle + \langle\psi(f), E_4(t)\psi(g)\rangle = 0 \text{ a.e.t } . \quad (5.1)$$

If f and g are continuous then (5.1) holds for all t. Varying f,g over continuous functions with compact support we obtain

$$\langle\psi(f), E_4(t)\psi(g)\rangle = 0 \quad \text{if} \quad f,g \quad \text{are continuous and supp } f,g \subset [0,t].$$

Since E_4 is adapted and $\{\psi(f), \text{ supp } f \subset [0,t], \text{ f continuous}\}$ is total in $H_t \otimes \Omega^t$ it follows that $E_4 = 0$ on $\mathcal{M}$. Varying f,g over continuous functions with $f(t_o) \neq 0$ and supp $g \subset [o,t_o)$ we conclude that $E_3(t_o) = 0$ on $\mathcal{M}$. Similarly $E_2(t_o) = 0$. Now varying f,g in (5.1) over continuous functions such that $f(t_o)g(t_o) \neq 0$ we obtain $E_1(t_o)=0$. Since t_o is arbitrary it follows that all the E_j vanish on $\mathcal{M}$. In other words we have shown that the four differentials $d\Lambda, dA, dA^\dagger, dt$ are linearly independent.

Now the question naturally arises whether there exist other independent differentials. We now try to answer this question.

If $dX = E_1 d\Lambda + E_2 dA + E_3 dA^\dagger$ then X is a martingale. Suppose that X, E_i $(1 \leq i \leq 3)$ are bounded operator valued processes such that

$$\max(||E_2(t)||^2, ||E_3(t)||^2) \leq \rho(t)$$

where ρ is locally summable. Then we obtain the inequality

$$\max(||X(t)u||^2 - ||X(s)u||^2, \; ||X^\dagger(t)u||^2 - ||X^\dagger(s)u||^2) \leq ||u||^2 \int_s^t \rho(\tau)d\tau \quad (5.2)$$

for all $u \in \Gamma(L_2[0,s]) \otimes \Omega^s$, $s < t$. In the light of this inequality we introduce the following definition.

<u>Definition 5.1</u> A bounded martingale X is called <u>regular</u> if there exists a Radon measure μ on $[0,\infty)$ such that for all $0 < s < t < \infty$

and $u \in \Gamma(L_2[0,s]) \otimes \Omega^s$

$$\max(||X(t)u||^2 - ||X(s)u||^2,\ ||X^\dagger(t)u||^2 - ||X^\dagger(s)u||^2)$$

$$\leq \ ||u||^2\ \mu((s,t]).$$

We are now ready to state the main result of this section.

Theorem 5.1 A bounded martingale X in $\Gamma(L_2[0,\infty))$ is regular if and only if there exist three bounded operator valued processes $\{E_i,\ 1 \leq i \leq 3\}$ such that the following conditions hold:

(1) $\max(||E_2(t)||^2,\ ||E_3(t)||^2) \leq \rho(t)$ a.e. t

for some locally summable function ρ in $[0,\infty)$;

(2) $dX = E_1\, d\Lambda + E_2\, dA + E_3\, dA^\dagger$.

Remarks A Hilbert-Schmidt martingale X in $\Gamma(L_2[0,\infty))$ is regular and it admits the representation

$$dX = -\ X\, d\Lambda + E\, dA + F\, dA^\dagger$$

where E and F are Hilbert-Schmidt adapted processes. A typical example of such a martingale is $\{|\psi(f_t)\rangle\langle\psi(g_t)|,\ t \geq 0\}$ where $|u\rangle\langle v|$ denotes the rank one operator defined by

$$|u\rangle\langle v|w = \langle v,w\rangle u, \qquad u,v,w \in \Gamma(L_2[0,\infty)).$$

Any unitary martingale X in $\Gamma(L_2[0,\infty))$ is regular. Furthermore, it admits the representation

$$dX = (K-1)X\, d\Lambda$$

where K is a unitary adapted process. If θ is a real Borel function on $[0,\infty)$, $\{\Gamma(e^{i\theta\chi_{[0,t]}}),\ t \geq 0\}$ is an example of a unitary martingale.

Theorem 5.1 is an unpublished result of the author and K.B. Sinha. Representations of Hilbert-Schmidt and unitary martingales were obtained in [18]. However, they follow from Theorem 5.1.

References

1. Accardi,L., Parthasarathy,K.R. : Stochastic calculus on local algebras. In : Quantum Probability and Applications, Accardi and von Waldenfels (ed.) (to appear).

2. Applebaum,D.B., Hudson,R.L. : Fermion Ito's formula and stochastic evolutions, Commun. Math. Phys. 96, 473-496 (1984).

3. Applebaum,D.B., Quasifree stochastic evolutions, In: Quantum Probability and Applications, Accardi and von Waldenfels (ed.) Lecture Notes in Mathematics, Berlin, Heidelberg, New York, Tokyo, Springer (to appear).

4. Araki,H. : Factorisable representations of current algebra, Publications of RIMS, Kyoto University, Ser. A 5, 361-422 (1970).

5. Barchielli,A., Lupieri,G : Quantum stochastic calculus, Operation valued stochastic processes and continual measurements in quantum theory, preprint, University of Milan.

6. Barnett,C., Streater, R.F., Wilde,I.F. : The Ito-Clifford integral I-IV, J. Functional Analysis, 48, 172-212(1982), J.London Math. Soc., 27, 373-384 (1983), Commun. Math. Phys., 89, 13-17 (1983), J. Operator Theory, 11, 255-271 (1984).

7. Barnett,C., Streater,R.F., Wilde,I.F. : Stochastic integrals in an arbitrary probability gage space, Math. Proc. Cambridge Phil. Soc., 94, 541-551 (1983).

8. Frigerio, A. : Diffusion processes, quantum stochastic differential equations and the classical KMS condition, J.Math. Phys. 25, 1050-65 (1984).

9. Frigerio, A., Gorini, V.: Markov dilations and quantum detailed balance, Commun. Math. Phys. 93, 517-32 (1984).

10. Gorini, V., Kossakowski, A., Sudarshan, E.C.G.: Completely positive dynamical semigroups of n-level system, J.Math. Phys. 17, 821-5 (1976).

11. Hudson, R.L., Lindsay, J.M.: Stochastic integration and a martingale representation theorem for non-Fock quantum Brownian motion, J. Functional Anal. 61, 202-221 (1985).

12. Hudson, R.L., Karandikar, R.L., Parthasarathy, K.R.: Towards a theory of noncommutative semimartingales adapted to

Brownian motion and a quantum Ito's formula; and Hudson, R.L., Parthasarathy,K.R. : Quantum diffusions. In : Theory and Applications of Random Fields, Kallianpur, (ed.) Lecture Notes in Control Theory and Information Sciences 49, Berlin, Heidelberg, New York, Tokyo : Springer 1983.

13. Hudson,R.L., Parthasarathy,K.R. : Quantum Ito's formula and stochastic evolutions, Commun. Math. Phys. 93, 301-323 (1984).

14. Hudson,R.L., Parthasarathy,K.R. : Stochastic dilations of uniformly continuous completely positive semigroups, Acta Appl. Math. 2, 353-398 (1984).

15. Hudson,R.L., Parthasarathy,K.R. : Generalised Weyl operators, In : Stochastic Analysis and Applications, Truman and Williams (ed.) Lecture Notes in Mathematics 1095, Berlin, Heidelberg, New York, Tokyo: Springer (1984).

16. Hudson,R.L., Parthasarathy,K.R. : Construction of quantum diffusions, In : Quantum Probability and Applications, Accardi (ed.) Lecture Notes in Mathematics 1055, Berlin, Heidelberg, New York, Tokyo: Springer (1984).

17. Hudson,R.L., Parthasarathy,K.R. : Unification of Fermion and Boson stochastic calculus, Submitted to Commun. Math. Phys.

18. Hudson,R.L., Lindsay,J.M., Parthasarathy,K.R.: Stochastic integral representations of some martingales in Fock space. Preprint, Nottingham University.

19. Lindblad,G. : On the generators of quantum dynamical semigroups, Commun. Math. Phys. 48, 119-30 (1976).

20. Parthasarathy,K.R., Schmidt,K. : Positive Definite Kernels, Continuous Tensor Products and Central Limit Theorems of Probability Theory, Lecture Notes in Mathematics 272, Berlin, Heidelberg, New York: Springer (1972).

21. Parthasarathy,K.R. : A remark on the integration of Schrödinger equation using quantum Ito's formula, Lett. Math. Phys. 8, 227-232 (1984).

22. Parthasarathy,K.R. : Some remarks on the integration of Schrödinger equation using the quantum stochastic calculus, In : Quantum Probability and Applications, Accardi and von Waldenfels (ed.) (to appear).

23. Parthasarathy,K.R. : Boson stochastic calculus, Pramana (to appear).

24. Parthasarathy,K.R. : One parameter semigroups of completely positive maps on groups arising from quantum stochastic differential equations, Bolletino Mat. Ital. (to appear).

25. Segal,I.E. : Tensor algebras over Hilbert spaces, Ann. Math. 63, 160-175 (1956).

26. Streater,R.F. : Current commutation relations, continuous tensor products and infinitely divisible group representations, Rendiconti Sc. Int. Fisica. Fermi, Vol. XI, 247-269 (1969).

27. von Waldenfels,W. : Ito solution of the linear quantum stochastic differential equation describing light emission and absorption, In : Quantum Probability and Applications, Accardi (ed.), Lecture Notes in Mathematics 1055, Berlin, Heidelberg, New York, Tokyo : Springer (1984).

QUANTUM THEORY AND STOCHASTIC PROCESSES - SOME CONTACT POINTS

L. STREIT
BiBoS
Fakultät für Physik
Universität Bielefeld
D-4800 Bielefeld 1, FRG

In recent years, progress in mathematical physics as well as in the theory of random processes and fields has frequently been inspired by a cross-disciplinary exchange of problems, methods and ideas.

Examples of such joint progress can be found in the research on Feynman and Wiener integrals, random Schroedinger operators, and Euclidean Field Theory. We sketch some of the development in these fields of physics and conclude with some remarks on mathematical methods and their application in physics in two short sections on Dirichlet Forms and on Hida's White Noise Calculus.

As a cautionary and apologetic note it is proper to add that a complete overview over that rapidly expanding subject would have gone way beyond the limitations of the present framework and the capabilities of the author. This article should best be considered as a first introduction to some (by far not all) of the literature.

- FEYNMAN AND WIENER INTEGRALS -

Quantum mechanics is a probabilistic theory and the structure of quantum dynamics is tantalizingly similar to that of (Markovian) stochastic processes. This structure finds a particularly compact and general expression in the "Feynman Integral" for expectation values of quantum mechanical observables

$$\langle A \rangle = \int \mathcal{D}^{\infty}[x] \;\; A[x]\, e^{\frac{i}{\hbar} S[x]} \quad .$$

For the physicist this integral representation is a powerful conceptual as well as computational tool, for a recent state of the art report see e.g. [GI 86]. For the mathematician, on the other hand, it is a priori meaningless, and, if he is so inclined, a challenge: to develop structures in the framework of which it can be defined, understood, and manipulated.

A general heading under which such attempts will fall is Infinite Dimensional Analysis. As an example I mention the framework of oscillatory integrals (Fresnel Integrals) in infinite dimension elaborated by Albeverio and Høegh-Krohn [AH 76] and then "manipulated" by J. Rezende [Re 84]. In particular, the stationary phase approximation to the Feynman integral generates the "semiclassical approximation" of quantum mechanics valid for $\hbar \to 0$.

At the present conference, approaches using analyticity arguments resp. the Dyson expansion have been presented by D.A. Storvick, K.S. Chang, and G.W. Johnson. It is desirable, and indeed possible, to give a direct measure theoretic meaning to the Feynman integral. For "actions" $S[\cdot]$ decomposable into a free and an "interaction" term

$$S = S_0 + S_I$$

with suitable properties,

$$\int \mathcal{D}^\infty[x] \; e^{\frac{i}{\hbar} S_I[x]}$$

can be interpreted as a complex measure on Poisson processes. This was first observed by Maslov and Chebotarev [MC 79], and later developed extensively by Blanchard, Combe, Høegh-Krohn, Sirugue, Sirugue-Collin, Rodriguez [BC 86], [CR 81]. In particular, this method is flexible enough to accommodate the description of spinning particles, as in [BC 85], [DJ 83].

Representations of Schroedinger theory directly in terms of Brownian motion have been given by Azencott and Doss [AD 84], and in the framework of generalized Brownian functionals by Hida and Streit [SH 82]. Both approaches offer a straightforward evaluation of the semiclassical limit $\hbar \to 0$.

The structural similarity

Feynman Integral ⟵⟶ Wiener Integral

and correspondingly

Schroedinger Equation ⟵⟶ Heat Equation

has been noticed long ago. More recently, a deeper and more powerful relation has become clear:

unitary time development group in quantum mechanics $U_t = e^{-\frac{i}{\hbar}Ht}$	Markovian semigroups $e^{-Ht}(x,y) = \rho_t(x,y)$

In the finite dimensional case $(x,y \in \mathbb{R}^d)$ this correspondence has deepened the insight into Quantum Dynamics, particularly for very singular interactions which could not be treated previously by conventional methods, we shall return to this point later. Here we mention as application only the treatment of quantum-mechanical multiwell (tunneling) problems using the Ventcel-Friedlin technique for the corresponding diffusion processes [Jo 80].

The stochastic reformulation has led within physics to the question whether there might be "more to it", i.e. whether a stochastic reinterpretation might be suggested.

Do the Markov processes generated by Schroedinger Hamiltonians have some kind of "physical reality(?)", are they maybe even "fundamental", in the sense that the quantum phenomena should arise as superficial effects of a deeper stochastic structure? The question [Ne 85] is not decided.

On the other hand, the large amount of work done by Albeverio et al. [AHS 77], Carlen [Ca 84], Guerra [Gu 83], Nelson [Ne 85] and many others on the interrelation of quantum and stochastic mechanics has led to many useful applications in the opposite direction. Stochastic equations of motion

$$dX_t = \beta(X_t)dt + \alpha dB_t$$

with β the "mean foreward velocity" and $\alpha(\sim \hbar^{1/2})$ the strength of the (quantum) "noise" dB_t is reformulated in terms of a Schroedinger type operator - a linearization of the problem. Invariant distributions are simply eigenstates of this operator, etc. We return to this point in detail later; here we mention the recent developments of a stochastic calculus of variations, such as in the work of Blanchard and Zengh [BZ 85] Zheng and Meyer [MZ 85], Guerra and Morato [GM 83], Yasue [Ya 81], and Zambrini [Za 84].

A remarkable sideproduct of this Hamiltonian reformulation is based on the introduction of auxiliary Grassmann variables ("supersymmetry"). With these techniques Luttinger [Lu 83] arrives at a heuristic but extremely effective rederivation of Donsker-Varadhan large deviation results, a derivation which in particular suggests a much more general validity of these results, see below and [FT 84].

- STOCHASTIC POTENTIALS -

Quite another contact point of quantum theory and stochastic processes is furnished by quantum Hamiltonians

$$H = H_o + V$$

where $V = V_\omega(x)$ is a (possibly generalized) random field, modelling forces of fluctuating strength and/or location. Such models are used to describe the properties of "random media", such as glasses, materials with impurities, suspensions, etc. Important physical questions concern the scattering of waves and particles in random media, resp. the localization of states, metallic versus insulator properties of materials; the density of states

$$N(E) = \lim_{\text{vol.}\to\infty} \frac{\#\{k: E_k \le k\}}{\text{vol.}}$$

should be calculated or at least estimated, etc. Best understood mathematically are models where the potential $V_\omega(x)$ is metrically transitive ("ergodic") under a discrete or continuous group of translations

$$x \to x + a \in \mathbb{R}^d , \quad \text{with} \quad a \in \mathbb{R}^d \quad \text{or} \quad a \in \mathbb{Z}^d ,$$

and here in particular the one-dimensional case $d = 1$. Among the results so far obtained are those on the spectrum of H_ω which turns out to be sure and non-discrete, as well as asymptotic, threshold, and certain regularity properties of the density of states $N(E)$. In the one-dimensional case one has further the Lyapunov index which is linked to the density of states by Thouless' formula [Th 72], [Is 73], [Pa 80]. It vanishes on intervals on which the spectrum of H_ω is absolutely continuous. Kotani [Ko 82] has shown that it is strictly positive for a class of "non-deterministic" potentials.

Related models arise if one discretizes not only the translation group but also the configuration space: $x \in \mathbb{Z}^d$ (Anderson model). The unperturbed Hamiltonian H_0 is then a bounded (difference) operator on the Hilbert space $\ell^2(\mathbb{Z}^d)$, a simplification which makes the Anderson model a useful tool to discuss "localization", the breakdown of diffusion in periodic systems (crystals) when the periodicity is perturbed randomly, at least for sufficiently large disorder and low energy. For recent results see e.g. [FM 85], [Im 85], [ST 85]. A recent review on random potentials is [KM 84]; for the structural similarities with the quantum theory of almost periodic potentials see e.g. [BL 85].

- QUANTUM FIELDS, STOCHASTIC FIELDS -

The central problem of Quantum Field theory, and in fact the cornerstone for a fundamental theory of elementary particles is the construction of <u>relativistic, interacting quantum fields</u>. In the framework of a stochastic reformulation of quantum theory this translates itself into the construction of certain

<u>Euclidean-invariant non-Gaussian random fields</u>.

The existence then of the corresponding quantum fields is guaranteed if the random fields obey the "Osterwalder-Schrader axioms" (see e.g. [Si 74]), in particular a (possibly weakened) Markov property.

A typical ansatz for the construction of Euclidean fields is

$$d\mu[\varphi] = \lim Z^{-1} e^{-S[\varphi]} d\mu_o[\varphi]$$

where S is a local functional of the field which specifies the interaction. μ_o usually is assumed Gaussian. The ansatz for its characteristic functional

$$\tilde{\mu}_o = e^{-\frac{1}{2}(f,(-\Delta + m^2)^{-1}f)}$$

corresponds to the (Euclidean resp. quantum free field of mass m. Note that the normalization constant

$$Z = \lim < e^{-S[\varphi]} >_{\mu_o}$$

has the form of a particle function of equilibrium statistical mechancis (in this language the Laplacean in the definition of μ_o induces a nearest neighbour coupling of ferromagnetic type).

It should be noted that, because of the sample function properties of μ_o multiplicative local functionals such as $e^{-S[\varphi]}$ require various regularizations of $\varphi(x)$, $x \in \mathbb{R}^d$, increasingly so for the space-time dimensions $d = 1,2,3,4,\ldots$: Our shorthand notation lim stands for the necessity of removing these regularizations, with μ defined as a weak limit of the regularized measures. Conventional regularizations lead to a Gaussian (i.e. trivial) limit at least for $d > 4$, while a rich structure of Euclidean and quantum fields have been constructed in this fashion for $d \leq 3$ (see e.g. [Gl 81] for a review). In the following we list some structural problems of current interest:

- A particular feature of Markov fields on unbounded domains, such as $\mathbb{Z}^d$ or $\mathbb{R}^d$, $d \geq 2$, is the strengthening of the Markov property to include boundaries of domains which are unbounded [AH 80], [Fö 80]. Recently, Röckner [Rö 85] and Zegarlinski [Ze 85] have investigated the correspondence between local specifications for (Gaussian or non-Gaussian) Markov fields and the Dirichlet problem for the corresponding (linear or nonlinear) PDE of classical Euclidean field theory, with the global Markov property, corresponding to uniqueness of solutions. (For Gaussian fields, Dynkin and Röckner have established the equivalent of global Markov property and locality [Dy 80], [Rö 85]).
- The feature of phase transitions is shared by Euclidean field theories and equilibrium statistical mechanics. The investigation of "critical points" (scale in-

variant or "automodel" random fields) is undertaken with renormalization group methods [GK 85], [It 85]. A particular remarkable recent application of renormalization group technqiues was the construction of a non-superrenormalizable field theory [GK 85] (i.e. one where an infinite number of subtractions is needed to make the action functional well-defined).
Another recent development in this context is the investigation with renormalization group methods of random fields not on $\mathbb{R}^d$ but on fractal domains such as the Sierpinski gasket [GM 83], [HH 85].

- In the expression given above for the characteristic function of a free Euclidean field one notes that the covariance kernel is the Green function of a diffusion process: of Brownian motion X_t with exponential killing m^2. In other words, the Hamiltonian of the Euclidean field φ is the Dirichlet form of the process X. Recently, E. Dynkin has undertaken a study of the "occupation field" t constructed from the diffusion X_t by

$$T_x = \int_0^{\xi} \delta(X_t - x)dt$$

where ξ is the death time of the process. The occupation field T is closely related to the second Hermite polynomial (Wick square) of the Euclidean field φ. Powers of T_x measure the returns of the Brownian path to the point x (self-intersections) thus reducing the study of polynomial interaction of fields to that of a gas of Brownian paths which interact where they touch as first proposed by Symanzik [Sy 69], [Va 69], [BF 82], [Ai 84].

- In this context Edwards' Brownian path model for interacting polymers has attracted renewed interest. It is characterized by a partition function

$$Z_T = \langle e^{-\int^T\int \delta(B_t - B_s)dtds} \rangle .$$

Interaction occurs if the Brownian path intersects itself: a continuum limit of a self-repellent random walk. Existence in d=2-dimensional space was established by Varadhan [Va 69] and for $d = 3$ by Westwaster [We 82]. Kusuoka has studied the $T \to \infty$ asymptotics for $d = 1$ and path properties for $d = 3$ [Ku 85]. Albeverio et al. [   ] have studied the crossing of two independent Brownian motions

$$Z = \langle e^{-\iint \delta(B_t^{(1)} - B_s^{(2)})dtds} \rangle.$$

Via the Feynman-Kac formula this corresponds to a heat equation with stochastic potential

$$V(x,B^{(2)}) = \int \delta(x - B_s^{(2)})ds .$$

- Considerable effort has gone into numerical "Monte Carlo" studies of non-Gaussian Gibbs measures such as the $d\mu[\varphi]$ above [La 80]. An initial configuration of the random field is constructed, and by random variations it is driven towards an (the?)

invariant distribution. Mathematically this amounts to the construction of quantum theories via the "stochastic quantization" method proposed by Parisi and Wu [PW 81]. Some recent results of this method were presented at this conference by Namiki et al., and by Potthoff. Apart from its proximity to numerical procedures, advantages are (1) its flexibility, exemplified by the discussion of "bottomless" (non-semi-bounded) actions" $S[\varphi]$ by [AH 83],[BP 84], (2) the manageability of constrained systems [NO 84,85]; [Se 84] also has a review of the subject, and (3) the structural insight gained from a "supersymmetric" formulation of stochastic quantization in terms of auxiliary Grassman variables [PS 82], [Ki 84]. Among the applications of such supersymmetric extensions of functional integration are e.g. the discussion of "dimension reduction" of Ising models in the presence of random magnetic fields [Ca 82] (note however [Im 85],[FF 84]). Another success of a supersymmetric reformulation was the extremely straightforward, albeit heuristic, derivation of Donsker-Varadhan large deviation results for the Wiener process by Luttinger [Lu 83], indicative of their validity for a much larger class of diffusion processes.

Apart from random paths, random surfaces play an increasingly important role in statistical mechanics (phase boundaries, surface phenomena) as well as in quantum field theory (string theory, surface representations of gauge theories). Literature concerning this topic can be found in [Fr 84], for Lie group which functions on random surfaces (Markov co-surfaces) and their application in gauge field theory see [AH 81].

- QUANTUM THEORY, DIRICHLET FORMS, AND DIFFUSIONS -

For nonrelativistic quantum mechanics the most elegant (and fruitful) contact with the theory of stochastic processes is through local Dirichlet forms ("energy forms")

$$\varepsilon(f) = \int_{\mathbb{R}^d} (\nabla f)^2 d\mu \ .$$

For quantum dynamics generated typically by a self-adjoint operator H which, at least formally, is given by

$$H = -\Delta + V(x)$$

and which has a state ψ associated with the infimum of its spectrum ("ground state") one has

$$(\varphi, H\varphi) = \varepsilon(f)$$

if $\varphi(x) = f(x)\psi(x)$ and $d\mu = \psi^2 dx$; the potential V is related to the ground state ψ by

$$V = \frac{\Delta\psi}{\psi} \ .$$

While this reformulation can be established rigorously for a wide class of perturbations [AHS 77] its merits become particularly obvious for such measures μ for which the equation

$$\varepsilon(f) = (\varphi, H\varphi)$$

defines a (self-adjoint, semibounded) quantum Hamiltonian, while the equation $V = \frac{\Delta\psi}{\psi}$ breaks down: this is the important case of interactions, so singular that they do not admit the usual description in terms of a perturbation V. Hence for the physics' side, the merit of the method lies in the many important examples of singular interactions which find a well-defined framework in "Dirichlet quantum mechanics" and can be handled effectively therein.

The investigation of energy forms, of the quantum theories and of the stochastic processes which they generate has in recent years been an excellent example of fruitful cross-fertilization between mathematics and physics research. For a recent review see [FU 85]; here we can only mention a few of those developments.

- *Approximation of processes and Hamiltonians*. It is of obvious interest to approximate the above-mentioned extremely singular perturbations by more regular ones. In the framework of energy forms the quest is for conditions that guarantee continuity of the Markovian generator as a function of the energy form; in physics terminology continuity of Schroedinger Hamiltonians as functions of their ground states. Criteria for (strong resolvent) convergence are given in [AHS 80], [AK 85]. Characteristically there are corresponding convergence results for the solutions of sequences of stochastic differential equations.

- *Stochastic mechanics* is related - as mentioned above - to quantum mechanics; and most directly so through energy forms: the mean foreward velocity is the logarithmic derivative of the probability density ψ^2 which defines the energy form.

- *Recurrence, transience, explosion* [AHS 77]. Here we mention only Ichihara's recent criterion for non-explosion: if $\psi^2 > 0$ on all compacts and

$$\psi^2 < a\, e^{b|x|} \quad x \in \mathbb{R}^d$$

then the probability for a finite killing time is zero [Ic 84]; note that no smoothness conditions for the invariant distribution ψ^2 are imposed.

- *Impenetrable barriers* for diffusions of arbitrary dimension spaces, and hence for dynamical systems of quantum theory as well as of stochastic mechanics are described in terms of separating regions of arbitrarily small capacity, with nodal surfaces

$$\{ x_; \ \psi^2(x) = 0 \}$$

as the simplest example [AF 81].

Applications are manifold: in biology for the spatial pattern formation of populations [Na 80], in atmospheric physics for zonal wind patterns [AB 85], in astrophysics for the regularities in planet formation (Bode-Titius law) [AB 83].

- *On manifolds* the energy forms

$$\varepsilon(f) = \int_M \langle df, df\rangle \, \rho(x) \sqrt{g} \, dx$$

correspond to perturbations of the Laplace-Beltrami operator Δ:

$$Hf = -\Delta f + \langle d \ln \rho \, , \, df\rangle \quad f \in C_0^\infty(M).$$

(For details see [Fu 85].) Physics provides many examples of diffusions on manifolds; spin relaxation of molecules can be described as a diffusion on the (group) manifold of molecular configurations [Ba 82].

- *Random ground states* $\psi = \psi_\omega(x)$ provide an interesting alternative to random potentials

$$V = V_\omega(x)$$

for the description of random media [Na 80].

- *Non-local Dirichlet forms* (which correspond to jump processes [AH 78]) offer a framework for the description of non-local interactions in quantum theory [Me 85].

- *Donsker-Varadhan asymptotics* can be extended from Brownian motion to the more general diffusions described by energy forms [FT 84].

- *On infinite-dimensional spaces* Dirichlet forms can be constructed in analogy to the finite dimensional case by considering e.g.

$$\varepsilon(F) = \int_{S^*} (\Delta F)^2 \, d\mu$$

for tame functionals on S^* of the form

$$F(f) = F(\langle f, e_1\rangle, \ldots, \langle f, e_n\rangle) \quad e_i \in S$$

and looking for closed extensions [AH 77]. Recently, Takeda has established the existence and uniqueness of a "Markovian" extension for the case where the measure μ is absolutely continuous with respect to the Wiener measure

$$d\mu = \rho \, d\mu_W$$

with a smooth Radon-Nikodym derivative

$$\rho'^{1/2} \in C^{\infty}$$

in the sense of Malliavin calculus.

A direct construction of (closed) Dirichlet forms in the infinite-dimensional case was undertaken by Kusuoka [Ku 82].

- *For energy forms on Lie group* - valued functions and their connection with the representation theory for infinite-dimensional Lie groups see e.g. [AH 81, 83].

- HIDA'S WHITE NOISE ANALYSIS -

Infinite-dimensional analysis is a natural habitat for the theory of (oscillatory or stochastic) path integrals; and the White Noise measure, characterized by its Fourier transform

$$\tilde{\mu}(f) = e^{-\frac{1}{2}(f,f)}$$

is the next best thing to the infinite-dimensional flat "measure" which physicists like to invoke under notations such as

$$\mathcal{D}^{\infty} x \quad \text{or} \quad \prod_{t \in T} dx(t) \text{ , etc.}$$

White Noise analysis has been developed extensively by T. Hida, and by T, Kubo and S. Takenaka; see [Hi 85] for a short review and references. At this conference, it is the subject of one of the main talks [Kuo 85], so I shall not present it systematically but restrict myself to some remarks to highlight its rich applicability.

- The White Noise measure μ is invariant under the transformations induced by orthogonal transformations of the test-functions f: $L^2(d\mu)$ carries a representation of the "infinite-dimensional rotation group" of such transformations with many interesting subgroups [Hi 85].
- The decomposition of $L^2(d\mu)$ into multiple Wiener integrals induces a Fockspace structure

  $$L^2(d\mu) = \bigoplus_{k=0}^{\infty} H_n .$$

  The "annihilation" operator

  $$\partial_t : H_n \to H_{n-1}$$

and its adjoint ∂_t^* serve to construct

- Representations of the Canonical Commutation Relations:

$$[\partial_t^* , \cdot \partial_s] = \delta(t - \delta).$$

- Stochastic integrals with respect to Brownian motion [Ku 83]:

$$\int F \, dB_t = \int dt \, \partial_t^* F(B_t) \; .$$

  The derivation of Ito's formula is straightforward in this framework.

- Generalizations of this calculus to non-abelian "quantum stochastic processes" as e.g. in [ST 84]. (Quantum probability is one of the growth points at the probability physics interface [AW 84].) We refer to the review at this conference by one of its pioneers [Pa 85].

- Various generalizations of the Laplacean to infinitely many dimensions [Kuo 85], such as e.g. the Beltrami Laplacean [Hi 83]

$$- \Delta_B = \int dt \, \partial_t^* \partial_t \quad .$$

(The physicist recognizes the "number operator".)

- (hence) as a starting point for Malliavin calculus [Po 85].

- Riggings

$$(L^2)_+ \subset L^2(d\mu) \subset (L^2)_-$$

  open up a wide field of infinite dimensional analysis and provide a framework for many examples of generalized Brownian (or White Noise) functionals, such as Donsker's δ-function, or Feynman integrals [SH 82], [Kuo 83].

In short, applications are many, and the mathematicians' proposition "Let us use White Noise" [Hi 78] will prompt the physicists' request: "Let us have White Noise Calculus".

REFERENCES

[AB 83] S. Albeverio, Ph. Blanchard, A stochastic model for orbits of planets and satellites, Expo. Math. 4, 365 (1983)

[AB 85] S. Albeverio, Ph. Blanchard, in: Quantum Probability and Application II, Lect. Not. in Math. 1136, Springer 1985

[AD 84] R. Azencott, H. Doss, L'équation de Schroedinger quand $\hbar$ tend vers zero; une approche probabiliste, ZiF Preprint, Bielefeld 1984

[AF 81] S. Albeverio, M. Fukushima, W. Karwowski, L. Streit, Capacity and Quantum Mechanical Tunneling, Comm. Math. Phys. 81, 501 (1981)

[AH 76] S. Albeverio, R. Høegh-Krohn, Mathematical theory of Feynman path integrals, Lect. in Math. 523, Springer 1976

[AH 77] S. Albeverio, R. Høegh-Krohn, Dirichlet Forms and Diffusion Processes on Rigged Hilbert Spaces, Z. Wahrsch. 40, 1 (1977)

[AH 78] S. Albeverio, R. Høegh-Krohn, The Structure of Diffusion Processes, Preprint 1978

[AH 80] S. Albeverio, R. Høegh-Krohn, Uniqueness and the Global Markov Property, in: Quantum Fields - Algebras, Processes, (L. Streit, ed.), Springer, Vienna 1980

[AH 81] S. Albeverio, R. Høegh-Krohn, and D. Testard, Irreducibility and Reducibility for the Energy Representation of the Group of Mappings of a Riemannian Manifold into a Compact Semisimple Lie Group, J. Funct. Anal. 41, 378 (1981)

[AH 83] S. Albeverio, R. Høegh-Krohn, D. Testard and A. Vershik, Factorial Representation of Path Groups, J. Funct. Anal. 51, 115 (1983)

[AH 84] S. Albeverio, R. Høegh-Krohn, H. Holden, Markov Cosurfaces and Gauge Fields, Schladming 84; Acta Phys. Austr. Suppl. XXVI, 211 (1984)

[AH 85] S. Albeverio, R. Høegh-Krohn, H. Holden, Stochastic Multiplicative Measures, Generalized Markov Semigroups and Groupvalued Stochastic Processes and Fields, BiBoS preprint, Bielefeld 1985, to appear in (BiBoS I)

[AHS 77] S. Albeverio, R. Høegh-Krohn, L. Streit, Energy Forms, Hamiltonians, and distorted Brownian Motion, J. Math. Phys. 18, 907 (1977)

[AHS 80] S. Albeverio, R. Høegh-Krohn, L. Streit, Regularization of Hamiltonians and Processes, J. Math. Phys. 21, 1636 (1980)

[Ai 84] M. Aizenman, The Intersection of Brownian Paths as a Case Study of a Renormalization Group Method for Quantum Field Theory, Comm. Math. Phys. 92, 19 (1984)

[AK 85] S. Albeverio, S. Kusuoka, L. Streit, Convergence of Dirichlet forms and associated Schroedinger operators, J. Funct. Anal. (to appear)

[AW 84] L. Accardi, W. v. Waldenfels (eds.), Quantum Probability and Applications II, Springer Lect. N. in Math. Nr. 1136, Berlin 1984

[BP 84] Ph. Blanchard, J. Potthoff, R. Sénéor, A Remark on Perturbation Expansions for Unstable Actions via Stochastic Quantization, BiBoS preprint 19, Bielefeld 1984

[BC 85] Ph. Blanchard, Ph. Combe, M. Sirugue, M. Sirugue-Collin, Probabilistic Solution of the Dirac Equation, BiBoS preprint 44, Bielefeld 1985

[BS 85] Ph. Blanchard, M. Sirugue, Large Deviations from Classical Paths. Hamiltonian flows as Classical Limits of Quantum Flows", Comm. Math. Phys. 101, 173 (1985)

[BZ 85] Ph. Blanchard, W. Zheng, Stochastic variational principle and diffusion processes, to appear in BiBoS preprints (1985)

[BC 86] Ph. Blanchard, Ph. Combe, M. Sirugue, M. Sirugue-Collin, Jump Processes in Quantum Theories, Proceedings Ascona-Como Conference: Stochastic processes in classical and quantum systems, Ascona - 24 - 49 - June 1985, to appear in Lect. Notes in Phys.

[Ba 82] M. Baldo,Brownian Motion in Group Manifolds: Application to Spin Relaxation of Molecules, Physica 114A, 88 (1982)

[BF 82] I.D. Brydges, J. Fröhlich, T. Spencer, The Random Walk Representation of Classical Spin Systems and Correlation Inequalities, Comm. Math. Phys. 83, 123 (1982)

[Ca 84] E. Carlen, Conservative Diffusion, Comm. Math. Phys. 94, 293 (1984)

[Ca 82] J. Cardy, Random Fields, Supersymmetry and Dimensionality Reduction, Workshop on Functional Integral Methods, Santa Barbara 1982

[CR 81] Ph. Combe, R. Høegh-Krohn, R. Rodriguez, M. Sirugue, M. Sirugue-Collin, Generalized Poissons Processes in Quantum Mechanics and Field Theory, Phys. Rep. 77, 221 (1981)

[DJ 83] G.F. De Angelis, G. Jona-Lasinio, M. Sirugue, Probabilistic Solution of the Pauli Equation, J. Phys. A16, 2433 (1983)

[Dy 80] E.B. Dynkin, Markov processes and random fields, Bull. Amer. Math. Soc. 3 (1980) 975

[Dy 83] E.B. Dynkin, Markov Processes as a tool in Field Theory, J. Funct. Anal. 50, 167 (1983)

[Dy 84] E.B. Dynkin, Gaussian and Non-Gaussian Random Fields Associated with Markov Processes, J. Funct. Anal. 55, 344 (1984)

[Dy 84a] E.B. Dynkin, Polynomials of the Occupation Field and Related Random Fields, J. Funct. Anal. 58, 20 (1984)

[Dy 85] E.B. Dynkin, Random Fields Associated with Multiple Points of the Brownian Motion, J. Funct. Anal. (1985)

[FF 84] D. Fisher, J. Fröhlich, T. Spencer, The Ising Model in a Random Magnetic Field, J. Stoch. Phys. 34, 863 (1984)

[Fö 80] H. Föllmer, On the Global Markov Property, in: Quantum Fields - Algebras, Processes, (L. Streit, ed.), Springer, Vienna 1980

[Fr 84] J. Fröhlich, Statistical Mechanics of Random Surfaces, Schladming 1984, Acta Phys. Austr. Suppl. XXVI, 255 (1984)

[FM 85] J. Fröhlich, F. Martinelli, E. Scoppola, T. Spencer, Constructive Proof of Localization in the Anderson Tight Binding Model, Comm. Math. Phys. 101, 1 (1985)

[Fu 80] M. Fukushima, Dirichlet Forms and Markov Processes, Kodansha-North Holland 1980

[FT 84] M. Fukushima, M. Takeda, A Transformation of a Symmetric Markov Process and the Donsker-Varadhan Theory, Osaka J. Math. 21, 311 (1984)

[Fu 85] M. Fukushima, Energy Forms and Diffusion Processes, in: Mathematics + Physics, Lectures on Recent Results, vol. I (L. Streit, ed.), World Scient. Publ. Singapore 1985

[GK 85] K. Gawędzki, A. Kupiainen, Exact Renormalization for the Gross-Neveu Model of Quantum Fields, Phys. Rev. Lett. 54. 2191 (1985)

[GH 83] J. Greensite, B. Halpern, Nucl. Phys. B211, 402 (1983)

[GL 81] J. Glimm, Quantum Physics, a fucntional integral point of view, Springer 1981

[GM 83] Y. Gefen, Y. Meir, B. Mandelbrot, A. Aharonov, Geometric Implementation of Hypercubic Lattices with Noninteger Dimensionality by Use of Low Lacunarity Fractal Lattices, Phys. Rev. Lett. 50, 145 (1983)

[Gu 83] F. Guerra, L. Morato, Quantization of Dynamical Systems and Stochastic Control Theory, Phys. Rev. D27, 1771 (1983)

[GI 86] M.C. Gutzwiller, A. Inomata, J.R. Klauder, L. Streit (Eds.), Path Integrals from meV to MeV, World Scient. Publ. Singapore, to appear 1986

[HH 85] K. Hattori, T. Hattori, H. Watanabe, Block Spin Approach to Fractal Field Theory.
New Approximate Renormalization Methods on Fractals,
Tokyo preprints UT-454, 457 (1985)

[Hi 78] T. Hida, Let us use White Noise, ZiF, Bielefeld 1978

[Hi 83] T. Hida, Generalized Brownian Functionals and Stochastic Integrals, Preprint (Baton Rouge 83)

[Hi 85] T. Hida, Brownian Functionals and the Rotation Group, in: Mathematics + Physics, Lectures on Recent Results, vol. I (L. Streit, ed.), World Scient. Publ. Singapore

[Ic 84] K. Ichihara, Explosion Problems for Symmetric Diffusion Processes", Proc. Japan Acad. 60 (1984)

[Im 85] J. Imbrie, The Ground state of the Three-Dimensional Random-Field Ising Model, Commun. Math. Phys. 98, 145 (1985)

[Is 73] K. Ishii, Localization of Eigenstates and Transport Phenomena in one-dimensional Disordered Systems, Supp. Theor. Phys. 53, 77 (1973)

[It 85] K. Ito, Contribution to this conference

[Jo 80] G. Jona-Lasinio, Stochastic Dynamics and the Semiclassical Limit of Quantum Mechanics, in: Quantum Fields - Algebra, Processes (L. Streit, ed.), Springer 1980

[Jo 85] G. Jona-Lasinio, F. Martinelli, E. Scoppola, Multiple tunnelings in d-dimension: a quantum particle in a hierarchical potential, Ann. Inst. H. Poincaré 42, 3 (1985)

[Ka 83] G. Kallianpur (ed.), Theory and Application of Random Fields, Springer LN in Control and Information Sciences, No. 49, Berlin 1983

[Ki 85] W. Kirsch, On a Class of Random Schroedinger Operators, Bochum preprint

[KM 85] W. Kirsch, F. Martinelli, Random Schroedinger Operators: Recent Results and Open Problems, in: Trends and Developments in the Eighties (S. Albeverio, Ph. Blanchard eds.), World Scientific Publ. Singapore 1985

[Ki 84] R. Kirschner, Stochastic Quantization and Supersymmetry, Schladming 84, Acta Phys. Austr. Suppl. XXVI, 409 (1984)

[Ko 82] S. Kotani, Lyapunov indices determine absolutely continuous spectra of stationary random one-dimensional Schroedinger operators, Proc. Kyoto Stoch. Conf. 1982

[Ku 83] I. Kubo, Ito Formula for generalized Brownian Functionals, in: [Ka 83], p. 156

[KT 80-82] I. Kubo, S. Takenaka, Calculus on Gaussian White Noise I-IV, Proc. Japan Acad, 56A, 376 (1980, 56A, 411 (1980), 57A, 433 (1981), 58A, 186 (1982)

[Kuo 83] H.-H. Kuo, Donsker's Delta Function as a Generalized Brownian Functional and its Application, in [Ka 83], p. 167

[Kuo 85] H.-H. Kuo, The Diffusion Process Associated with Lévy's Laplacean, Inv. talk, at this conference

[Ku 82] S. Kusuoka, Dirichlet Forms and Diffusion Processes on Banach Spaces, J. Fac. Sci. Univ. Tokyo 29, 79 (1982)

[Ku 85] S. Kusuoka, Asymptotics of Polymer Measures in One Dimension, On the Path Property of Edwards' Model for Long Polymer Chains in Three Dimensions, in: Infinite dimensional analysis and stochastic processes (S. Albeverio, ed.) Pitman, Boston 1985

[La 86] Chr. Lang, Computer Quantum Field Theory, in: Mathematics + Physics, Lectures on Recent Results, vol. II (L. Streit, ed.) World Scientific Publ. Singapore 1986

[Lu 83] J.M. Luttinger, A new method for the asymptotic evaluation of a class of path integrals, J. Math. Phys. 23, 1011 (1982), The asymptotic evaluation of a class of path integrals II, J. Math. Phys. 24, 2070 (1983)

[MC 79] V.P. Maslov, A.M. Chebotarev, Processus à saut et leurs applications dans la mécanique quantique, Springer Lect. Notes in Math. No. 106, 1979

[MZ 85] P. Meyer, Zheng Weian, Sur la construction de certaines diffusions, to appear in: Séminaire de Probabilité XX, Lect. Not. Math. 1985

[Na 80] M. Nagasawa, Segregation of Population in an Environment, J. Math. Biology 9, 213 (1980)

[NO 84] M. Namiki, I. Ohba, K. Okana, Stochastic Quantization of Constrained Systems - General Theory and Nonlinear σ-Model, Prog. Theor. Phys. 72, 350 (1984)

[No 85] M. Namiki, I. Ohba, K. Okano, M. Rikihasa, S. Tanaka, Numerical Simulation of Non-Linear σ-Model by Means of Stochastic Quantization Methods, Prog. Theor. Phys. 73, 186 (1985)

[Ne 85] E. Nelson, Quantum Fluctuations, Princeton University Press 1985

[PW 81] G. Parisi, Y.S. Wu, Perturbation Theory without Gauge Fixing, Scient. Sinic. 24, 483 (1981)

[PS 82] G. Parisi, N. Sourlas, Supersymmetric Field Theories and Stochastic Differential Equation , Nucl. Phys. B206, 321 (1982)

[Pa 80] L. Pastur, Spectral Properties of Disordered Systems in the One-Body Approximation, Comm. Math. Phys. 75, 179 (1980)

[Re 84] J. Rezende, Stationary Phase Method on Hilbert Space and Semi-Classical Approximation in Quantum Mechanics, ZiF preprint, Bielefeld 1984

[Re 84] J. Rezende, The Method of Stationary Phase for Oscillatory Integrals on Hilbert Spaces, Bielefeld preprint Bi-TP 1984/3

[Rö 85] M. Röckner, A Dirichlet problem for distributions and specifications for random fields, Mem. AMS. 1985

[Rö II 85] M. Röckner, Generalized Markov fields and Dirichlet forms, Acta Appl. Math. 3, 285 (1985)

[Se 84] E. Seiler, Stochastic Quantization and Gauge Fixing in Gauge Theories, Acta Phys. Austr. Suppl. XXVI, 259 (1984)

[SZ 84] E. Seiler, D. Zwanziger, Nucl. Phys. B239, 177 (1984)

[Si 74] B. Simon, The φ(2) Euclidean field theory, Princeton Univ. Press 1974

[ST 85] B. Simon, M. Taylor, T. Wolff, Some Rigorous Results for the Anderson Model, Phys. Rev. Lett. 54, 1589 (1985)

[SH 82] L. Streit, T. Hida, White Noise Analysis and its Application to Feynman Integral, in: Measure Theory and its Applications (J.M. Belley et al., eds.), Springer LNM 1033, Berlin 1982

[Sy 69] K. Symanzik, Euclidean Quantum Field Theory, in: Local Quantum Theory (R. Jost, ed.), Academic Press, N.Y. 1969

[Ta 84] M. Takeda, On the Uniqueness of Markovian Extensions of Diffusion Operators on Infinite Dimensional Spaces (Osaka Math. J.), ZiF preprint, Bielefeld 1984

[Th 72] D. Thouless, A Relation between the Density of States and Range of Localization for One-Dimensional Random Systems, J. Phys. C5, 77 (1972)

[Va 69] R.S. Varadhan, (Appendix to [Sy 69].)

[Me 85] R. Vilela Mendes, Reconstruction of Dynamics from an Eigenstate, J. Math. Phys., in print

[Wa 83] S. Watanabe, Malliavin's Calculus in Terms of Generalized Wiener Functionals, in: [Ka 83] , p. 284

[We 82] J. Westwater, On Edwards' Model for Polymer Chains III. - Borel Summability, Comm. Math. Phys. 84, 459 (1982)

[Ya 81] K. Yasue, Stochastic Calculus of Variations, J. Funct. Anal. 41, 327 (1981)

[Za 84] J. Zambrini, Stochastic Dynamics: A Review of Stochastic Calculus of Variations, Princeton preprint 1984

[Ze 85] B. Zegarlinski, The Gibbs Measures and Partial Differential Equations,
Part I Ideas and Local Aspects
Part II The Global Aspects
BiBoS preprint nr. 25 (1985)

[Ze 84] B. Zegarlinski, Uniqueness and the Global Markov Property for Euclidean Fields: The Case of General Exponential Interactions, Comm. Math. Phys. 96, 195 (1984)

THE USE OF PACKING MEASURE IN THE ANALYSIS OF RANDOM SETS

S. James Taylor
Department of Mathematics
University of Virginia
Charlottesville, Va. 22903 USA

The scientific community have become increasingly aware of 'fractals' to a large extent because of the books of B. Mandelbrot [7] with their beautiful computer simulations of random sets. Fractal sets arise naturally in models appropriate for describing phenomena in a wide range of disciplines from astronomy to turbulence, biology, chemistry, or even economics. Scientists have now learned to recognise a fractal, though there is no clear consensus about a precise definition. In fact, there are several distinct procedures leading to a fractal index or dimension for sets without interior, so it would seem wise to reserve the term fractal for sets whose index is the same when calculated by any of these methods; but one would also want to exclude sets which look like pieces of a topological manifold. Fractals arise naturally from the sample paths of a stochastic process: in the present paper we only consider the random sets which are trajectories of a Lévy process. These sets and their generalizations have received a lot of attention in the past (see Adler [1]): we want to look at them afresh using the new machinery of packing measure as defined in a recent paper with Tricot [11]. This procedure yields precise measures in $\mathbb{R}^n$ which are invariant under any transformation preserving the metric and the measures then lead to a new definition of dimension.

The fractal measure most studied by probabilists is Hausdorff measure (see Falconer [3]) which is defined using economical covers of the set $E \subset \mathbb{R}^d$ by small diameter sets. Let us recall the spherical version of this definition. Start with a monotone $\phi:(0,1)\to(0,1)$ such that ϕ is right continuous and $\phi(0+)=0$, and for which there is a finite constant K with

$$\phi(2s)/\phi(s) \le K \quad \text{whenever} \quad 0<s<\tfrac{1}{2}. \tag{1}$$

Condition (1) is a weak smoothness condition: clearly $\phi(s) = s^\alpha$ for $\alpha>0$ satisfies all these properties. Now define the set function, for $E \subset \mathbb{R}^d$,

$$\phi\text{-}m(E) = \lim_{\delta\downarrow 0} \inf \{\sum_{i=1}^{\infty} \phi(2r_i) : E \subset \bigcup_{i=1}^{\infty} B_{r_i}(x_i),\ r_i<\delta\} \tag{2}$$

where $B_{r_i}(x_i)$ denotes an open ball of radius r_i and center x_i and the infimum in (2) is taken over all covers of E by balls of radius $< \delta$.

ϕ-m is a metric outer measure on the subsets of $\mathbb{R}^d$, so the class of ϕ-measurable sets includes the Borel sets. The value ϕ-m(E) can be zero, finite and positive, infinite σ-finite or non σ-finite. By specialising to $\phi(s) = s^\alpha$, $\alpha > 0$ we obtain the fractal index

$$\dim E = \inf\{\alpha > 0 : s^\alpha\text{-m}(E) = 0\} \tag{3}$$

usually called the Hausdorff-Besicovitch dimension of E.

To obtain packing measure we try first to take the mirror image of the definition (2) by seeking to 'pack' as many disjoint balls as possible on the set. We require the centers of the balls to lie in the set E. Consider

$$\phi\text{-P}(E) = \lim_{\delta\downarrow 0} \text{suf}\,\{\sum_{i=1}^{\infty} \phi(2r_i) : B_{r_i}(x_i) \text{ disjoint, } x_i \in E, r_i < \delta\} \tag{4}$$

The condition $x_i \in E$ could be weakened, but the discussion in [11] shows that we cannot replace it by the requirement $E \cap B_{r_i}(x_i) \neq \Phi$. Since $\phi\text{-P}(\bar{E}) = \phi\text{-P}(E)$ we see that ϕ-p is not an outer measure, by taking for E a countable dense set in $\mathbb{R}^d$. Even if we look only at compact sets the difficulty is still there: let E_0 be the countable set $\{0, n^{-1}$ for $n \in N\}$ and take the obvious packing of balls radius $(n+1)^{-2}$ centered at n^{-1} to give $s^{\frac{1}{2}}\text{-P}(E_0) = +\infty$ while E_0 is the countable union of singletons $\{x_i\}$ and $s^{\frac{1}{2}}\text{-P}\{x_i\} = 0$. However, the set function ϕ-P satisfies all the conditions for a pre-measure so we can obtain an outer measure by defining

$$\phi\text{-p}(E) = \inf\{\sum_{i=1}^{\infty} \phi\text{-P}(E_i) : E \subset \bigcup_{i=1}^{\infty} E_i\} \tag{5}$$

This set function ϕ-p is a metric outer measure with nice topological properties. We call ϕ-p(E) the ϕ-packing measure of E, and analogous to (3) we define the fractal index

$$\text{Dim } E = \inf\{\alpha > 0 : s^\alpha\text{-p}(E) = 0\} \tag{6}$$

For all sets E it was shown in [11] that $\phi\text{-m}(E) \leq \phi\text{-p}(E)$, and it is easy to see that $s^\alpha\text{-m}(\mathbb{R}^d) = 0$, for $\alpha > d$ so that, for all subsets of $\mathbb{R}^d$ we have

$$0 \leq \dim E \leq \text{Dim } E \leq d. \tag{7}$$

The two dimension indices are distinct: for each $0 \leq \alpha \leq \beta \leq d$ we can define a compact $E_0 \subset \mathbb{R}^d$ with $\dim E_0 = \alpha$, $\text{Dim } E_0 = \beta$. However, the upper and lower entropy indices (see Hawkes [5]) lie between dim and Dim so that the condition $\dim E_0 = \text{Dim } E_0$ forces all the usual fractal indices to take the same value. The two-stage definition (4),(5) seems complicated, but proofs are simplified by the fact that one can always approximate ϕ-p(E) by ϕ-P(F) for suitable subsets $F \subset E$ in the sense that

$$\phi\text{-p}(E) = \inf \{\lim \phi\text{-P}(F_n) : F_n \uparrow E\}.$$

In [11] we considered the trajectory of a Brownian path B(t) in $\mathbb{R}^d$ and it follows that, for $d \geq 2$,

$$\dim B[0,1] = \text{Dim } B[0,1] = 2 \quad \text{a.s.}$$

In fact, in the transient case $d \geq 3$, we have a precise packing measure function $\psi(s) = s^2/\log\log \frac{1}{s}$ such that finite positive constants c_d exist with

$$\psi\text{-p}(B[0,1]) = c_d \quad \text{a.s.} \tag{8}$$

This brings me to my first

PROBLEM 1. *Decide whether or not there is a correct packing measure function for planar Brownian motion. If there is none, characterise those measure functions $\phi(s)$ which give zero or infinite ϕ-p measure*

I conjecture that, for B(t) in $\mathbb{R}^2$ and a fixed measure function ϕ

$$\phi\text{-p } B[0,1] = 0 \quad \text{a.s.} \quad \text{or} \quad \phi\text{-p } B[0,1] = +\infty \quad \text{a.s.},$$

in which case there is no analogue to (8). I now want to present two new results, a precise one for a strictly stable process in $\mathbb{R}^d$ of index $\alpha < d$, and a dimension result for a general Lévy process. If X(t) is stable of index α it is easy to show that

$$\dim X[0,1] = \text{Dim } X[0,1] = \min(\alpha, d) \quad \text{a.s.},$$

but instead we obtain the more precise result

THEOREM 2. *Suppose $h(s) = s^\alpha \psi(s)$ where ψ is monotone increasing and satisfies (1), and X is a strictly stable process in $\mathbb{R}^d$ of index $\alpha < d$. Then*

$$h\text{-p}X[0,1] = \begin{matrix} 0 \text{ a.s.} \\ +\infty \text{ a.s.} \end{matrix} \quad \text{according as} \quad \int_{0+} \frac{[\psi(s)]^2}{s}\, ds \quad \begin{matrix} < +\infty \\ = +\infty \end{matrix} .$$

Before we can prove this we need to recall from [11] two technical lemmas which are analogous to the corresponding results for Hausdorff measures. The density theorem result is a mirror image of that for Hausdorff measure (theorem 5.4).

LEMMA 3. *For each ϕ satisfying (1) there is a finite λ such that for all $E \subset \mathbb{R}^d$, Borel measures μ with $0 < ||\mu|| = \mu(\mathbb{R}^d) < +\infty$,*

$$\lambda\mu(E)\ \inf_{x\in E}\{\limsup_{r\downarrow 0}\frac{\phi(2r)}{\mu(B_r(x))}\} \le \phi\text{-}p(E)$$

$$\le ||\mu|| \sup_{x\in E}\{\limsup_{r\downarrow 0}\frac{\phi(2r)}{\mu(B_r(x))}\}.$$

Now let Γ be the class of semi-dyadic cubes in $\mathbb{R}^d$: $C \in \Gamma$ if for some integer n, C is a cube of side length 2^{-n} and each of its projections $\text{proj}_i\, C$ on the ith axis is a half open interval of the form $[\frac{1}{2}k_i 2^{-n}, (\frac{1}{2}k_i+1)2^{-n})$ with $k_i \in \mathbb{Z}$. Each x in $\mathbb{R}^d$ belongs to 2^d cubes of Γ with side 2^{-n}; of these we denote by $v_n(x)$ the unique cube in Γ of side 2^{-n} whose complement is at distance 2^{-n-2} from the dyadic cube of side 2^{-n-2} which contains x. This forces all the coordinates of x to be within 2^{-n-2} of the corresponding co-ordinates for the centre of $v_n(x)$. Now put

$$\Gamma_E = \{v_n(x) : n \in \mathbb{N},\ x \in E\}$$

and use cubes from Γ_E to replace the balls $B_r(x)$ of definition (4), with diameter $C = d^{\frac{1}{2}}2^{-n}$ replacing $2r$ = diameter $B_r(x)$. For any ϕ satisfying (1) this gives a new packing pre-measure ϕ-P^{**} which is comparable to ϕ-P premeasure in the sense that there are constants C_1, C_2 such that, for all Borel sets $E \subset \mathbb{R}^d$,

$$c_1\phi\text{-}P(E) \le \phi\text{-}P^{**}(E) \le c_2\phi\text{-}P(E)$$

If we apply the final step (5) of the definition to ϕ-P^{**} we get a packing measure ϕ-p^{**} with the same class of sets of zero measure or finite measure.

LEMMA 4. *For each ϕ satisfying (1) there are finite positive constants c_1, c_2 such that for all $E \subset \mathbb{R}^d$,*

$$c_1\phi\text{-}p(E) \le \phi\text{-}p^{**}(E) \le c_2\phi\text{-}p(E)$$

Now use the sample path to define a random measure

$$\mu(E) = |\{t \in [0,1] : X(t) \in E\}|. \tag{9}$$

This gives a Borel measure with $||\mu|| = 1$, and μ is concentrated on $X[0,1]$ and spread evenly on it. If $x = X(t_0)$, $0 < t_0 < 1$ then $\mu(B_r(x))$ is the total time spent by $X(s)$ in the ball $B_r(x)$ up to time 1. Considered as a random process for small positive r it is clearly equivalent to the process $T(r) = T_1(r) + T_2(r)$ where

$$T_1(r) = \int_0^\infty I_{B_r(0)}X(s)ds$$

is the total time spent by X in $B_r(0)$ and $T_2(r)$ is the corresponding sojourn time for an independent copy of the dual X^1 of X obtained by time reversal. Note that if X is strictly stable of index α, so is X^1. We need the lower asymptotic growth rate of $T(r)$.

LEMMA 5. For $h(s)$ as in theorem 2, we have

$$\liminf_{r\downarrow 0} \frac{T_1(r)+T_2(r)}{h(2r)} = \begin{matrix} 0 \text{ a.s.} \\ +\infty \text{ a.s.} \end{matrix} \quad \text{according as} \int_{0+} \frac{[\psi(s)]^2}{s}\, ds \begin{matrix} = +\infty \\ < +\infty \end{matrix}.$$

Proof. We use the notation $f(r) \approx g(r)$ as $r \downarrow 0$ to mean that there are constants $c_1, c_2, \delta > 0$ such that $c_1 f(r) \le g(r) \le c_2 f(r)$ for $0 < r < \delta$. First suppose $\int_{0+} \frac{[\psi(s)]^2}{s}\, ds$ converges. For any fixed λ define

$$E_{\lambda,n} = \{w: T_1(2^{-n}) + T_2(2^{-n}) < \lambda h(2^{-n+1})\}$$

$$\cap\, D_{1,\lambda,n} \sim D_{2,\lambda,n} \quad \text{where}$$

$$D_{i,\lambda,n} = \{w: T_i(2^{-n}) < \lambda h(2^{-n+1})\}.$$

Since $D_{1,\lambda,n}, D_{2,\lambda,n}$ are independent and

$$P(D_{i,\lambda,n}) < P\{w: |X_i(\lambda h(2^{-n+1}))| > 2^{-n}\}$$

$$\approx \psi(2^{-n}), \text{ by scaling, we get}$$

$$\sum P(E_{\lambda,n}) \le c \sum [\psi(2^{-n})]^2$$

which is a convergent series. Thus $E_{\lambda,n}$ happens finitely often a.s. for each λ, and hence $\lim_{r\downarrow 0} \frac{T_1(r)+T_2(r)}{h(2r)} = +\infty$ a.s.

In the other direction we use $D_{1,\lambda,n} \cap D_{2,\lambda,n} \supset E_{2\lambda,n}$ but replace $D_{i,\lambda,n}$ by a smaller event

$$F^k_{i,\lambda,n} = \{w: |X_i(\lambda h(2^{-n}))| < k(\lambda h(2^{-n}))^{1/\alpha},$$

$$|X_i(\lambda h(2^{-n+1})) - X_i(\lambda h(2^{-n}))| > (2k)2^{-n},$$

$$X_i(s) \text{ does not enter } B_{2^{-n}}(0) \text{ for } s > \lambda h(2^{-n+1})\}$$

where k is a large positive integer. Using the Markov property allows us to obtain

$$P(F^k_{i,\lambda,n}) \ge \tau_k P(H^k_{i,\lambda,n}), \text{ where}$$

$$H^k_{i,\lambda,n} = \{w: |X_i(\lambda h(2^{-n+1})) - X_i(\lambda h(2^{-n}))| > 2k2^{-n}\}$$

and $\tau_k \to 1$ as $k \to \infty$. For each fixed k, if $\int_{0+} \frac{[\psi(s)]^2}{s} ds = +\infty$, we now get

$$\sum P(F^k_{1,\lambda,n} \cap F^k_{2,\lambda,n}) = +\infty \text{ and, for } n \neq m,$$

$$P(F^k_{i,\lambda,n} \cap F^k_{i,\lambda,m}) \leq P(H^k_{i,\lambda,n} \cap H^k_{i,\lambda,m})$$

$$\leq \tau_k^{-2} P(F^k_{i,\lambda,n}) P(F^k_{i,\lambda,m}).$$

A standard version of Borel-Cantelli now tells us that $F^k_{1,\lambda,n} \cap F^k_{2,\lambda,}$ happens infinitely often with probability at least τ_k^4. Hence for each λ, $D_{i,\lambda,n} \cap D_{2,\lambda,n}$ and therefore $E_{2,\lambda,n}$ happens infinitely often with probability at least τ_k^4. Let $k \to \infty$ to give $E_{2,\lambda,n}$ happens infinitely often a.s. for each $\lambda \to 0$ which in turn implies that $\liminf_{r\downarrow 0} \frac{T_1(r)+T_2(r)}{h(2r)} = 0$ a.s.

<u>Proof of Theorem 2.</u> Whenever $\int_{0+} \frac{[\psi(s)]^2}{s} ds = +\infty$ we now know that, if μ is the measure given by (9) for each fixed $t_0 \in (0,1)$ a.s.

$$\liminf \frac{\mu B_r(X(t_0))}{h(2r)} = 0$$

or

$$\limsup_{r\downarrow 0} \frac{h(2r)}{\mu B_r(x(t_0))} = +\infty \tag{10}$$

Now let F be the subset of values $t_0 \in (0,1)$ satisfying (10). A Fubini argument tells us that $|F| = 1$ so that $\mu X(F) = 1$ a.s. The density Lemma 3 now implies that $\phi\text{-p}X(F) = +\infty$, so that $\phi\text{-p}X[0,1] = +\infty$ a.s.

As with corresponding Hausdorff measure calculations the use of the density theorem in the other direction only yields, under the condition $\int_{0+} \frac{[\psi(s)]^2}{s} ds < \infty +$, that $\phi\text{-p}X(G) = 0$ where

$$G = \{t_0 \in (0,1): \liminf_{r\downarrow 0} \frac{\mu B_r(X)t_0))}{h(2r)} = +\infty\}$$

is the set of 'good' points. Consider

$$Q_n = \{t_o \in (0,1): \liminf_{r\downarrow 0} \frac{\mu B_r(X(t_0)}{h(2r)} \leq n\}$$

For fixed n we consider

$$E\, h\text{-p}^{**}X(Q_n) \leq E\, h\text{-P}^{**}X(Q_n).$$

Now we can only get a contribution to $h\text{-}P^{**}X(Q_n)$ from semi-dyadic cubes of side 2^{-k} if $X(s)$ hits the inside dyadic cube of side 2^{-k-2} and then starting from this hitting place it leaves the ball of radius 2^{-k-2} in less than $nh(2^{-k})$. The expected number of dyadic cubes of side 2^{-k-2} hit in $(0,1)$ is $O(2^{k\alpha})$ and the probability of being bad, given that it is hit as $O(\psi(2^{-k}))$. Leaving out the disjointness requirement gives

$$\begin{aligned} E\, h\text{-}P^{**}X(Q_n) &= O\Big(\sum_{k=k_0}^{\infty} E(N_k)h(2^{-k})\Big) \qquad (11)\\ &= O\Big(\sum_{k=k_0}^{\infty} 2^{k\alpha}h(2^{-k})\psi(2^{-k})\Big)\\ &= O\Big(\sum_{k_0}^{\infty} [\psi(2^{-k})]^2\\ &= 0, \text{ by letting } k_0 \to \infty, \end{aligned}$$

where N_k in (11) is the total number of semi-dyadic cubes of side 2^k whose centre is hit but from which there is a quick exit. It now follows that $h\text{-}P^{**}X(Q_n) = 0$ a.s., so $h\text{-}p^{**}X(\bigcup_{n=1}^{\infty} Q_n) = 0$ a.s. Since $G \cup \bigcup_{n=1}^{\infty} Q_n = (0,1)$ we have proved that $h\text{-}p^{**}X[0,1] = 0$ and we are finished by Lemma 4.

To obtain a result as precise as Theorem 2 for a general Lévy process seems hopeless, but in collaboration with Fristedt [4] we are able to obtain the analogous dichotomy for most subordinators.

<u>Packing dimension</u>. For each Lévy process X, Pruitt [8] defined an index

$$\gamma = \sup\{\alpha > 0 : \lim_{a\downarrow 0} a^{-\alpha} ET(a,1) = 0\}$$

where $T(a,1)$ is the occupation time of a ball radius a up to time 1, and proved that

$$\dim X[0,1] = \gamma \quad \text{a.s.}$$

Hendricks [6], as part of his work on uniform dimension results, considered the index

$$\gamma' = \sup\{\alpha > 0 : \liminf_{a\downarrow 0} a^{-\alpha} ET(a,1) = 0\} \qquad (12)$$

In [10] we consider the properties of γ' and show that

$$\tfrac{1}{2}\beta \leq \gamma' \leq \min(d,\beta)$$

where β is the upper index of Blumenthal Getoor. We also construct examples of a Levy process in $\mathbb{R}^d (d \geq 2)$ for which any prescribed values satisfying $0 \leq \gamma \leq \gamma' \leq 2$ are obtained and we give other equivalent definitions of γ'. For the present let us content ourselves with the main result:

THEOREM 6. If X is a Lévy process in $\mathbb{R}^d$, and γ' is the index defined by (12), then the packing dimension of the trajectory satisfies

$$\operatorname{Dim} X[0,1] = \gamma' \quad \text{a.s.}$$

Proof First suppose $\alpha < \gamma'$. Then, since γ' has the same value for the dual process X^1 we have

$$\liminf_{r \downarrow 0} \; E \frac{\mu B_r(X(t_0))}{r^\alpha} \leq \liminf_{r \downarrow 0} E \frac{(T_1(r,1)+T_2(r,1))}{r^\alpha}$$

$$= 0.$$

By Fatou's lemma, this implies that

$$\liminf \frac{\mu B_r(X(t_0))}{r^\alpha} = 0 \quad \text{a.s.}$$

for each fixed $t_0 \in (0,1)$. An application of Lemma 3 now gives $s^\alpha\text{-}pX[0,1] = +\infty$ a.s.

Now suppose $\alpha > \gamma'$. Neither Fatou nor the density Lemma now help, so we use a first moment argument directly. Suppose $\alpha > \beta > \gamma'$, then $r^{-\beta} E\, T_1(r) \to +\infty$ as $r \downarrow 0$. The semi-dyadic cubes of side 2^{-k} cover $\mathbb{R}^d$ 2^d times so, if N_k is the number of such cubes hit by $X[0,1]$ we have, by Lemma 5.1 of [9]

$$E(N_k) = o(2^{k\beta}) \quad \text{as} \quad k \to \infty.$$

Hence the expected contribution to $s^\alpha\text{-}P^{**}X[0,1]$ from cubes of side 2^{-k} is $O(2^{(\beta-\alpha)k})$. If $\delta = 2^{-k_0}$ this gives

$$Es^\alpha\text{-}P^{**}_\delta X[0,1] = O\Big(\sum_{k=k_0}^{\infty} (2^{(\beta-\alpha)k})\Big) \to 0 \quad \text{as} \quad k_0 \to \infty.$$

Hence $s^\alpha\text{-}P^{**}X[0,1] = 0$ a.s. By Lemma 4, $s^\alpha\text{-}pX[0,1] = 0$ a.s. and therefore $\operatorname{Dim} X[0,1] \leq \alpha$ a.s.

Taking sequences $\alpha_n \uparrow \gamma'$ and $\alpha_n \downarrow \gamma'$ completes the argument that

$$\operatorname{Dim} X[0,1] = \gamma' \quad \text{a.s.}$$

Our first attempts at evaluating packing measure for random sets lead us to believe that in most situations, there will be no exact measure function ϕ leading to a finite positive ϕ-p measure. Intuitively this is caused by the fact that occasional large jumps or gaps in the set allow you to pack another ball of that size. We close by asking whether only Lévy processes which have a finite Lévy measure have an exact packing measure function.

PROBLEM 7. For which Lévy processes X with infinite Lévy measure and no Brownian component can we find a measure function for which $0 < \phi\text{-}pX[0,1] < +\infty$ *a.s.*

Candidates to try for a positive solution to problem 7 would be processes with a very dense Lévy measure such as those considered by Fristedt [3].

REFERENCES

[1] R. J. Adler, The geometry of random fields, Wiley, 1980

[2] K. J. Falconer, The geomtry of fractal sets, Cambridge Press, 1985.

[3] B. B. Fristedt, Upper functions for symmetric processes with stationary independent movements, Indiana Math J. 21 (1971), 177-185.

[4] B. F. Fristedt and S. J. Taylor, The exact packing measure for the trajectory of a subordinator (in preparation).

[5] J. Hawkes, Hausdorff measure, entropy, and the independence of small sets, Proc. Lon. Math. Soc. 28 (1974), 700-724.

[6] W. J. Hendricks, A uniform lower bound for Hausdorff dimension for transient symmetric Lévy processes, Annals of Prob. 11 (1984), 589-592.

[7] B. B. Mandelbrot, The fractal geometry of nature, Freeman, 1982.

[8] W. E. Pruitt, The Hausdorff dimension of the range of a process with stationary independent increments. Jour. Math. and Mechanics 19 (1969), 371-378.

[9] W. E. Pruitt and S. J. Taylor, Sample path properties of processes with stable components, Z. Wahrscheinlichkeits-theorie, 12 (1969), 267-289.

[10] W. E. Puritt and S. J. Taylor, The packing dimension of the trajectory of a Lévy process (in preparation).

[11] S. J. Taylor and C. Tricot, Packing measure and its evaluation for a Brownian path, Trans. Amer. Math. Soc. 288 (1985), 679-699.

The author acknowledges financial support from National Science Foundation grant number DMS-8317815 for the period when the research described above was carried out.

Vol. 1062: J. Jost, Harmonic Maps Between Surfaces. X, 133 pages. 1984.

Vol. 1063: Orienting Polymers. Proceedings, 1983. Edited by J.L. Ericksen. VII, 166 pages. 1984.

Vol. 1064: Probability Measures on Groups VII. Proceedings, 1983. Edited by H. Heyer. X, 588 pages. 1984.

Vol. 1065: A. Cuyt, Padé Approximants for Operators: Theory and Applications. IX, 138 pages. 1984.

Vol. 1066: Numerical Analysis. Proceedings, 1983. Edited by D.F. Griffiths. XI, 275 pages. 1984.

Vol. 1067: Yasuo Okuyama, Absolute Summability of Fourier Series and Orthogonal Series. VI, 118 pages. 1984.

Vol. 1068: Number Theory, Noordwijkerhout 1983. Proceedings. Edited by H. Jager. V, 296 pages. 1984.

Vol. 1069: M. Kreck, Bordism of Diffeomorphisms and Related Topics. III, 144 pages. 1984.

Vol. 1070: Interpolation Spaces and Allied Topics in Analysis. Proceedings, 1983. Edited by M. Cwikel and J. Peetre. III, 239 pages. 1984.

Vol. 1071: Padé Approximation and its Applications, Bad Honnef 1983. Prodeedings. Edited by H. Werner and H.J. Bünger. VI, 264 pages. 1984.

Vol. 1072: F. Rothe, Global Solutions of Reaction-Diffusion Systems. V, 216 pages. 1984.

Vol. 1073: Graph Theory, Singapore 1983. Proceedings. Edited by K.M. Koh and H.P. Yap. XIII, 335 pages. 1984.

Vol. 1074: E.W. Stredulinsky, Weighted Inequalities and Degenerate Elliptic Partial Differential Equations. III, 143 pages. 1984.

Vol. 1075: H. Majima, Asymptotic Analysis for Integrable Connections with Irregular Singular Points. IX, 159 pages. 1984.

Vol. 1076: Infinite-Dimensional Systems. Proceedings, 1983. Edited by F. Kappel and W. Schappacher. VII, 278 pages. 1984.

Vol. 1077: Lie Group Representations III. Proceedings, 1982–1983. Edited by R. Herb, R. Johnson, R. Lipsman, J. Rosenberg. XI, 454 pages. 1984.

Vol. 1078: A.J.E.M. Janssen, P. van der Steen, Integration Theory. V, 224 pages. 1984.

Vol. 1079: W. Ruppert. Compact Semitopological Semigroups: An Intrinsic Theory. V, 260 pages. 1984

Vol. 1080: Probability Theory on Vector Spaces III. Proceedings, 1983. Edited by D. Szynal and A. Weron. V, 373 pages. 1984.

Vol. 1081: D. Benson, Modular Representation Theory: New Trends and Methods. XI, 231 pages. 1984.

Vol. 1082: C.-G. Schmidt, Arithmetik Abelscher Varietäten mit komplexer Multiplikation. X, 96 Seiten. 1984.

Vol. 1083: D. Bump, Automorphic Forms on GL (3,IR). XI, 184 pages. 1984.

Vol. 1084: D. Kletzing, Structure and Representations of Q-Groups. VI, 290 pages. 1984.

Vol. 1085: G.K. Immink, Asymptotics of Analytic Difference Equations. V, 134 pages. 1984.

Vol. 1086: Sensitivity of Functionals with Applications to Engineering Sciences. Proceedings, 1983. Edited by V. Komkov. V, 130 pages. 1984

Vol. 1087: W. Narkiewicz, Uniform Distribution of Sequences of Integers in Residue Classes. VIII, 125 pages. 1984.

Vol. 1088: A.V. Kakosyan, L.B. Klebanov, J.A. Melamed, Characterization of Distributions by the Method of Intensively Monotone Operators. X, 175 pages. 1984.

Vol. 1089: Measure Theory, Oberwolfach 1983. Proceedings. Edited by D. Kölzow and D. Maharam-Stone. XIII, 327 pages. 1984.

Vol. 1090: Differential Geometry of Submanifolds. Proceedings, 1984. Edited by K. Kenmotsu. VI, 132 pages. 1984.

Vol. 1091: Multifunctions and Integrands. Proceedings, 1983. Edited by G. Salinetti. V, 234 pages. 1984.

Vol. 1092: Complete Intersections. Seminar, 1983. Edited by S. Greco and R. Strano. VII, 299 pages. 1984.

Vol. 1093: A. Prestel, Lectures on Formally Real Fields. XI, 125 pages. 1984.

Vol. 1094: Analyse Complexe. Proceedings, 1983. Edité par E. Amar, R. Gay et Nguyen Thanh Van. IX, 184 pages. 1984.

Vol. 1095: Stochastic Analysis and Applications. Proceedings, 1983. Edited by A. Truman and D. Williams. V, 199 pages. 1984.

Vol. 1096: Théorie du Potentiel. Proceedings, 1983. Edité par G. Mokobodzki et D. Pinchon. IX, 601 pages. 1984.

Vol. 1097: R.M. Dudley, H. Kunita, F. Ledrappier, École d'Éte de Probabilités de Saint-Flour XII – 1982. Edité par P.L. Hennequin. X, 396 pages. 1984.

Vol. 1098: Groups – Korea 1983. Proceedings. Edited by A.C. Kim and B.H. Neumann. VII, 183 pages. 1984.

Vol. 1099: C.M. Ringel, Tame Algebras and Integral Quadratic Forms. XIII, 376 pages. 1984.

Vol. 1100: V. Ivrii, Precise Spectral Asymptotics for Elliptic Operators Acting in Fiberings over Manifolds with Boundary. V, 237 pages. 1984.

Vol. 1101: V. Cossart, J. Giraud, U. Orbanz, Resolution of Surface Singularities. Seminar. VII, 132 pages. 1984.

Vol. 1102: A. Verona, Stratified Mappings – Structure and Triangulability. IX, 160 pages. 1984.

Vol. 1103: Models and Sets. Proceedings, Logic Colloquium, 1983, Part I. Edited by G.H. Müller and M.M. Richter. VIII, 484 pages. 1984.

Vol. 1104: Computation and Proof Theory. Proceedings, Logic Colloquium, 1983, Part II. Edited by M.M. Richter, E. Börger, W. Oberschelp, B. Schinzel and W. Thomas. VIII, 475 pages. 1984.

Vol. 1105: Rational Approximation and Interpolation. Proceedings, 1983. Edited by P.R. Graves-Morris, E.B. Saff and R.S. Varga. XII, 528 pages. 1984.

Vol. 1106: C.T. Chong, Techniques of Admissible Recursion Theory. IX, 214 pages. 1984.

Vol. 1107: Nonlinear Analysis and Optimization. Proceedings, 1982. Edited by C. Vinti. V, 224 pages. 1984.

Vol. 1108: Global Analysis – Studies and Applications I. Edited by Yu.G. Borisovich and Yu.E. Gliklikh. V, 301 pages. 1984.

Vol. 1109: Stochastic Aspects of Classical and Quantum Systems. Proceedings, 1983. Edited by S. Albeverio, P. Combe and M. Sirugue-Collin. IX, 227 pages. 1985.

Vol. 1110: R. Jajte, Strong Limit Theorems in Non-Commutative Probability. VI, 152 pages. 1985.

Vol. 1111: Arbeitstagung Bonn 1984. Proceedings. Edited by F. Hirzebruch, J. Schwermer and S. Suter. V, 481 pages. 1985.

Vol. 1112: Products of Conjugacy Classes in Groups. Edited by Z. Arad and M. Herzog. V, 244 pages. 1985.

Vol. 1113: P. Antosik, C. Swartz, Matrix Methods in Analysis. IV, 114 pages. 1985.

Vol. 1114: Zahlentheoretische Analysis. Seminar. Herausgegeben von E. Hlawka. V, 157 Seiten. 1985.

Vol. 1115: J. Moulin Ollagnier, Ergodic Theory and Statistical Mechanics. VI, 147 pages. 1985.

Vol. 1116: S. Stolz, Hochzusammenhängende Mannigfaltigkeiten und ihre Ränder. XXIII, 134 Seiten. 1985.

Vol. 1117: D.J. Aldous, J.A. Ibragimov, J. Jacod, Ecole d'Été de Probabilités de Saint-Flour XIII – 1983. Édité par P.L. Hennequin. IX, 409 pages. 1985.

Vol. 1118: Grossissements de filtrations: exemples et applications. Seminaire, 1982/83. Edité par Th. Jeulin et M. Yor. V, 315 pages. 1985.

Vol. 1119: Recent Mathematical Methods in Dynamic Programming. Proceedings, 1984. Edited by I. Capuzzo Dolcetta, W.H. Fleming and T. Zolezzi. VI, 202 pages. 1985.

Vol. 1120: K. Jarosz, Perturbations of Banach Algebras. V, 118 pages. 1985.

Vol. 1121: Singularities and Constructive Methods for Their Treatment. Proceedings, 1983. Edited by P. Grisvard, W. Wendland and J.R. Whiteman. IX, 346 pages. 1985.

Vol. 1122: Number Theory. Proceedings, 1984. Edited by K. Alladi. VII, 217 pages. 1985.

Vol. 1123: Séminaire de Probabilités XIX 1983/84. Proceedings. Edité par J. Azéma et M. Yor. IV, 504 pages. 1985.

Vol. 1124: Algebraic Geometry, Sitges (Barcelona) 1983. Proceedings. Edited by E. Casas-Alvero, G.E. Welters and S. Xambó-Descamps. XI, 416 pages. 1985.

Vol. 1125: Dynamical Systems and Bifurcations. Proceedings, 1984. Edited by B.L.J. Braaksma, H.W. Broer and F. Takens. V, 129 pages. 1985.

Vol. 1126: Algebraic and Geometric Topology. Proceedings, 1983. Edited by A. Ranicki, N. Levitt and F. Quinn. V, 523 pages. 1985.

Vol. 1127: Numerical Methods in Fluid Dynamics. Seminar. Edited by F. Brezzi, VII, 333 pages. 1985.

Vol. 1128: J. Elschner, Singular Ordinary Differential Operators and Pseudodifferential Equations. 200 pages. 1985.

Vol. 1129: Numerical Analysis, Lancaster 1984. Proceedings. Edited by P.R. Turner. XIV, 179 pages. 1985.

Vol. 1130: Methods in Mathematical Logic. Proceedings, 1983. Edited by C.A. Di Prisco. VII, 407 pages. 1985.

Vol. 1131: K. Sundaresan, S. Swaminathan, Geometry and Nonlinear Analysis in Banach Spaces. III, 116 pages. 1985.

Vol. 1132: Operator Algebras and their Connections with Topology and Ergodic Theory. Proceedings, 1983. Edited by H. Araki, C.C. Moore, Ş. Strătilă and C. Voiculescu. VI, 594 pages. 1985.

Vol. 1133: K.C. Kiwiel, Methods of Descent for Nondifferentiable Optimization. VI, 362 pages. 1985.

Vol. 1134: G.P. Galdi, S. Rionero, Weighted Energy Methods in Fluid Dynamics and Elasticity. VII, 126 pages. 1985.

Vol. 1135: Number Theory, New York 1983–84. Seminar. Edited by D.V. Chudnovsky, G.V. Chudnovsky, H. Cohn and M.B. Nathanson. V, 283 pages. 1985.

Vol. 1136: Quantum Probability and Applications II. Proceedings, 1984. Edited by L. Accardi and W. von Waldenfels. VI, 534 pages. 1985.

Vol. 1137: Xiao G., Surfaces fibrées en courbes de genre deux. IX, 103 pages. 1985.

Vol. 1138: A. Ocneanu, Actions of Discrete Amenable Groups on von Neumann Algebras. V, 115 pages. 1985.

Vol. 1139: Differential Geometric Methods in Mathematical Physics. Proceedings, 1983. Edited by H. D. Doebner and J. D. Hennig. VI, 337 pages. 1985.

Vol. 1140: S. Donkin, Rational Representations of Algebraic Groups. VII, 254 pages. 1985.

Vol. 1141: Recursion Theory Week. Proceedings, 1984. Edited by H.-D. Ebbinghaus, G.H. Müller and G.E. Sacks. IX, 418 pages. 1985.

Vol. 1142: Orders and their Applications. Proceedings, 1984. Edited by I. Reiner and K. W. Roggenkamp. X, 306 pages. 1985.

Vol. 1143: A. Krieg, Modular Forms on Half-Spaces of Quaternions. XIII, 203 pages. 1985.

Vol. 1144: Knot Theory and Manifolds. Proceedings, 1983. Edited by D. Rolfsen. V, 163 pages. 1985.

...ler, Choquet Order and Simplices. VI, 143 pages. 1985.

Vol. 1146: Séminaire d'Algèbre Paul Dubreil et Marie-Paule Malliavin. Proceedings, 1983–1984. Edité par M.-P. Malliavin. IV, 420 pages. 1985.

Vol. 1147: M. Wschebor, Surfaces Aléatoires. VII, 111 pages. 1985.

Vol. 1148: Mark A. Kon, Probability Distributions in Quantum Statistical Mechanics. V, 121 pages. 1985.

Vol. 1149: Universal Algebra and Lattice Theory. Proceedings, 1984. Edited by S. D. Comer. VI, 282 pages. 1985.

Vol. 1150: B. Kawohl, Rearrangements and Convexity of Level Sets in PDE. V, 136 pages. 1985.

Vol 1151: Ordinary and Partial Differential Equations. Proceedings, 1984. Edited by B.D. Sleeman and R.J. Jarvis. XIV, 357 pages. 1985.

Vol. 1152: H. Widom, Asymptotic Expansions for Pseudodifferential Operators on Bounded Domains. V, 150 pages. 1985.

Vol. 1153: Probability in Banach Spaces V. Proceedings, 1984. Edited by A. Beck, R. Dudley, M. Hahn, J. Kuelbs and M. Marcus. VI, 457 pages. 1985.

Vol. 1154: D.S. Naidu, A.K. Rao, Singular Pertubation Analysis of Discrete Control Systems. IX, 195 pages. 1985.

Vol. 1155: Stability Problems for Stochastic Models. Proceedings, 1984. Edited by V.V. Kalashnikov and V.M. Zolotarev. VI, 447 pages. 1985.

Vol. 1156: Global Differential Geometry and Global Analysis 1984. Proceedings, 1984. Edited by D. Ferus, R.B. Gardner, S. Helgason and U. Simon. V, 339 pages. 1985.

Vol. 1157: H. Levine, Classifying Immersions into $\mathbb{R}^4$ over Stable Maps of 3-Manifolds into $\mathbb{R}^2$. V, 163 pages. 1985.

Vol. 1158: Stochastic Processes – Mathematics and Physics. Proceedings, 1984. Edited by S. Albeverio, Ph. Blanchard and L. Streit. VI, 230 pages. 1986.

Vol. 1159: Schrödinger Operators, Como 1984. Seminar. Edited by S. Graffi. VIII, 272 pages. 1986.

Vol. 1160: J.-C. van der Meer, The Hamiltonian Hopf Bifurcation. VI, 115 pages. 1985.

Vol. 1161: Harmonic Mappings and Minimal Immersions, Montecatini 1984. Seminar. Edited by E. Giusti. VII, 285 pages. 1985.

Vol. 1162: S.J.L. van Eijndhoven, J. de Graaf, Trajectory Spaces, Generalized Functions and Unbounded Operators. IV, 272 pages. 1985.

Vol. 1163: Iteration Theory and its Functional Equations. Proceedings, 1984. Edited by R. Liedl, L. Reich and Gy. Targonski. VIII, 231 pages. 1985.

Vol. 1164: M. Meschiari, J.H. Rawnsley, S. Salamon, Geometry Seminar "Luigi Bianchi" II – 1984. Edited by E. Vesentini. VI, 224 pages. 1985.

Vol. 1165: Seminar on Deformations. Proceedings, 1982/84. Edited by J. Ławrynowicz. IX, 331 pages. 1985.

Vol. 1166: Banach Spaces. Proceedings, 1984. Edited by N. Kalton and E. Saab. VI, 199 pages. 1985.

Vol. 1167: Geometry and Topology. Proceedings, 1985. Edited by J. Alexander and J. Harer. VI, 292 pages. 1985.

Vol. 1168: S.S. Agaian, Hadamard Matrices and their Applications. III, 227 pages. 1985.

Vol. 1169: W.A. Light, E.W. Cheney, Approximation Theory in Tensor Product Spaces. VII, 157 pages. 1985.

Vol. 1170: B.S. Thomson, Real Functions. VII, 229 pages. 1985.

Vol. 1171: Polynômes Orthogonaux et Applications. Proceedings, 1984. Edité par C. Brezinski, A. Draux, A.P. Magnus, P. Maroni et A. Ronveaux. XXXVII, 584 pages. 1985.

Vol. 1172: Algebraic Topology, Göttingen 1984. Proceedings. Edited by L. Smith. VI, 209 pages. 1985.